SETTING ASIDE ALL AUTHORITY

Setting Aside All Authority

Giovanni Battista Riccioli and the Science
against Copernicus in the Age of Galileo

INCLUDING THE FIRST ENGLISH TRANSLATION OF

MONSIGNOR FRANCESCO INGOLI'S

essay to Galileo disputing the

Copernican system,

and the first English translation of

RICCIOLI'S REPORTS

regarding his experiments with falling bodies and with the
effect of air resistance on falling bodies.

CHRISTOPHER M. GRANEY

University of Notre Dame Press

Notre Dame, Indiana

Copyright © 2015 by the University of Notre Dame
Notre Dame, Indiana 46556
www.undpress.nd.edu
All Rights Reserved

Manufactured in the United States of America

Library of Congress Cataloging-in-Publication Data

Graney, Christopher M., 1966–
    Setting aside all authority : Giovanni Battista Riccioli and the science
against Copernicus in the age of Galileo / Christopher M. Graney.
        pages   cm
    Includes bibliographical references and index.
    ISBN 978-0-268-02988-3 (paperback : alkaline paper)—
    ISBN 0-268-02988-1 (paperback : alkaline paper)
        1. Riccioli, Giovanni Battista, 1598–1671.   2. Astronomers—Italy—
Biography.   3. Jesuit scientists—Italy—Biography.   4. Copernicus,
Nicolaus, 1473–1543.   5. Astronomy—Italy—History—17th century.
6. Science—Italy—History—17th century.   I. Title.
    QB36.R386G73   2015
    523.2—dc23
                                                        2014047949

To my sister,

Laura Kathleen Graney (1971–2013),

who had a great affection for the night sky,

and who took great pride in her brother's work.

Contents

Illustrations and Tables

Acknowledgments

This book would not exist if not for the general support, and very specific help, of my wife, Christina Louise Knochelmann Graney. Many, many thanks go to her. She has been urging me to consider writing a book since as early as 2010. She has also lent me her language skills. All Latin translation in this book, and in my other writings, is the result of a team effort. In this team effort, I work out the science content that the writers are trying to communicate, while she works out the grammar and insists that we have the patience to pay attention to details, so that our translation accurately reflects what the writers say, as closely as possible. Tina has had minimal formal training in Latin, but language talent runs deep in her family. Once when visiting her mother, Louise Knochelmann, we took some Latin with us. We were struggling through a sentence, when Louise piped up from across the room with the solution to the problem. "When did you last study Latin?" I asked. "High school," came the answer, with a slight hint of "well, duh!" Louise was in her late 70s at the time. My son John Henry Graney is an Arabic linguist who also knows Spanish and Chinese, and is a lover of grammar. My son Joe Graney loves science, history, and argument. Such a family is valuable for this sort of work.

This work had its genesis in two things—a student question and a 2004 article by Leos Ondra in *Sky & Telescope* magazine. The student question was how people could look at the evidence a telescope provided, yet not accept the Copernican system. Numerous people in my astronomy classes or visiting my college's observatory have asked this question in various ways, but one asked in a particularly memorable way: How, he said, can I look at the sky and see that it is blue, but accept some guy telling me to believe it to be pink, because that is what is in the Bible? The question is about rejecting clear scientific evidence. Students at Jefferson Community & Technical College in my home town of Louisville, Kentucky, are often so bright and so talented that they stun me. Yet some are so poorly educated (I have had a student who had never heard of protons, neutrons, and electrons; another did not know that stars are more distant than planets) that I am equally stunned. Quite a few are deeply religious, in a very traditional manner. Very few feel an obligation simply to accept what academic authority tells them. Thus I had to have a *really* good answer to this question, if I wanted all my students to hear it, understand it, and not tune me out as just another head talking at them about how smart people like me have enlightened ignorant people like them. The only way I could see to do this is history. That is why this book is here. So I thank my students.

Ondra's article pointed me on the path to answering the student question, but the path itself was made passable by the history of astronomy community. The article was about Ondra's discovering in Galileo's notes that Galileo had observed double stars. But Galileo's double star observations, measurements, and calculations seemed at odds with the Copernican system. Here was clear scientific evidence *not* in favor of Copernicus. In time, this would lead to my learning more about the problems that observations of stars created for Copernicans. Key to this was help from the history of astronomy community, especially members of the listserv HASTRO-L and participants at the Notre Dame Biennial History of Astronomy Workshops. I did not know where to go with the "star size problem" beyond Galileo, and so two key events for me were Thony Christie and Harald Siebert suggesting I look at Simon Marius and Riccioli, respectively. I thank them: their comments led to breakthroughs. I owe thanks to many scholars who have offered help and constructive criticism, but must especially thank Mike Crowe, Owen Gingerich, Dennis Danielson, Yaakov Zik, Todd Timberlake, and Matt Dowd (my editor at the University of Notre Dame Press, and also the moderator of HASTRO-L and an organizer of the Notre Dame workshops). The history of astronomy community is one in which established, accomplished scholars are willing to offer to the newcomer blunt but constructive criticism while welcoming the newcomer's ideas. This community enabled an astronomer from a community college in Kentucky to establish a record of scholarly publication in the history of astronomy and now to write this book about the history of astronomy.

Of course, this book would not have been written without resources. Thanks to the internet, a community college astronomer has access to volume upon volume of rare books *at his desk*. I thank the History of Science Collections, University of Oklahoma Libraries; ETH-Bibliothek Zürich, Alte und Seltene Drucke; Google books; and Archive.org for providing me with a fantastic library of digitized copies of old books. I also thank the Louisville Free Public Library and the Jefferson Community & Technical College libraries, whose electronic resources I used much.

I thank Jefferson Community & Technical College. I have been a member of the Jefferson faculty since 1990. Jefferson has provided for me the classic academic life in which a project like this can take root and grow. Where possible, and despite constant budgetary pressures and cuts, Jefferson has always provided support for my work.

There are other acknowledgments, and more thanks owed, but this will have to suffice.

1 Giovanni Battista Riccioli
and the *New Almagest*

The *New Almagest* was astronomy. What was not in the *New Almagest* did not need to be known. England's top astronomer—the first Astronomer Royal, John Flamsteed—used the *New Almagest* as a textbook for public lectures at Gresham College in 1665. Indeed, anyone carrying a copy of the *New Almagest* to a lecture would have looked quite learned, for the book was printed in two volumes, each the size of one of today's large coffee-table books. Within those volumes were over fifteen hundred pages, filled with dense text and diagrams. Clearly, anyone who read the *New Almagest*, which was published in 1651 by an Italian astronomer by the name of Giovanni Battista Riccioli (1598–1671), knew something about astronomy.

Information on every conceivable topic in seventeenth-century astronomy could be found in the pages of the *New Almagest*. There were chapters devoted to the motions of celestial bodies, to discussions of geometry, and to representations of the appearance of Jupiter, Venus, and other planets as seen through the best telescopes. There were reports on, and tables of data from, physics experiments of different sorts, such as those involving bodies falling through air and through water. There were lots of tables of astronomical data. There was a highly detailed, full-spread map of the Moon, which featured an interesting new naming system for lunar features. On the map one lunar crater was named "Copernicus." Another was named "Galileo." A large smooth area was named the "Sea of Tranquility."

These names would stick—when the Apollo 11 lander the "Eagle" touched down on the moon in 1969, it landed in the "Sea of Tranquility," and Neil Armstrong would radio, "Houston, Tranquility Base here. The Eagle has landed."

Inside each volume of the *New Almagest* was an impressive frontispiece that represented the state of astronomical knowledge in the mid-seventeenth century (fig. 1.1). The frontispiece showed Jupiter having four moons, as Galileo had reported in his groundbreaking *Starry Messenger* of 1610. But the frontispiece also showed Jupiter having cloud bands[1]—a discovery made since Galileo. It showed Venus having a crescent phase, as Galileo had first observed with his telescope. It also showed Mercury with a crescent phase—another discovery made since Galileo. Of course it featured a cratered moon, Mars and Saturn, and more. The *New Almagest* frontispiece presented all the key discoveries that had been revealed by the telescope since Galileo had turned one to the heavens in 1609 (fig. 1.2).

Yet the main feature of the frontispiece was a set of scales, being held by Urania, the Greek muse of astronomy, while the many-eyed Argus, holding a telescope, looked on (fig. 1.3). The modern reader who studies the frontispiece will immediately recognize, on the left side of the scales, a representation of the world system of Nicolaus Copernicus (1473–1543)—the "heliocentric hypothesis." There is the Sun, in the middle of everything, with planets circling around it, and the Moon circling the third planet from the Sun.

On the right side of the scales is *not* a representation of the "geocentric hypothesis," the ancient world system of Ptolemy (who wrote the original *Almagest*) and Aristotle, in which Earth is in the center of everything, and everything circles around Earth. Galileo's famous *Dialogue* of 1632 concerned "the two chief world systems: Ptolemaic and Copernican." But in the *New Almagest*, the Ptolemaic system is not what is being weighed against the Copernican system. In Riccioli's frontispiece, the Ptolemaic world system is sitting off to the side, on the ground, looking a little forlorn. Clearly the telescopic discoveries shown above the scales, especially the phases of Venus and Mercury (which proved those bodies circle the Sun), have removed that ancient system from Riccioli's consideration.

Rather, on the right side of the scales is a representation of a world system that is definitely geocentric. There is the Earth, right in the center,

Figure 1.1. Frontispiece of Giovanni Battista Riccioli's 1651 *New Almagest*. Image courtesy History of Science Collections, University of Oklahoma Libraries.

Figure 1.2. Details from the *New Almagest* frontispiece, showing Jupiter with moons and cloud bands (left), and showing Mercury and Venus each with a crescent phase (right). Images courtesy History of Science Collections, University of Oklahoma Libraries.

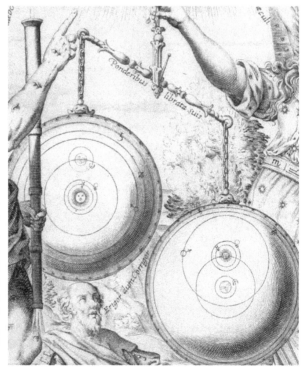

Figure 1.3. Detail from the *New Almagest* frontispiece, showing the world systems being weighed in the balance. Image courtesy History of Science Collections, University of Oklahoma Libraries.

with the Sun and Moon circling it. But this world system is not purely geocentric. Planets circle the Sun. This world system may be geocentric, but it has some heliocentric features. It is a "hybrid geocentric" (or "geo-heliocentric") hypothesis.

Moreover, the balance is obviously tipping in favor of this hybrid geocentric world system. The *New Almagest* is clearly promoting the view that, in the age of new telescopic discoveries, this hybrid geocentrism is superior to Copernican heliocentrism. And yet the hand of God reaches down from above the page, indicating "Numerus," "Mensura," "Pondus" ("Number," "Measure," "Weight"), while verses from the Vulgate Bible[2] grace the picture:

> "dies diei eructat verbum et nox nocti indicat scientiam"—"Day to day uttereth speech, and night to night sheweth knowledge" (Psalm 18:3)

> "videbo caelos tuos; opera digitorum tuorum"—"I will behold thy heavens, the works of thy fingers" (Psalm 8:4)

> "non inclinabitur in saeculum saeculi"—"[Earth] shall not be moved for ever and ever" (Psalm 103:5)

What is Riccioli saying, anyway, by representing the latest discoveries with all this?

Riccioli, a Jesuit priest who was originally trained as a theologian, before he discovered and fell in love with astronomy,[3] appears to be saying that, in light of telescopic discoveries, in light of a serious scientific analysis (numbering, measuring, weighing), Copernicus is wrong. The Earth does not move. Indeed, beyond the frontispiece, a very large part of the *New Almagest* consists of an analysis of the world system debate, and this analysis concludes with a question and a main conclusion. The question is this: according purely to reason, and setting aside all authority, which hypothesis may be asserted as true—that which supposes the motion of the Earth, or that which supposes the immobility of the Earth?[4] And what was the conclusion?

> Reasoning and intrinsic arguments alone considered, and every authority set aside; the hypothesis supposing the immobility or quiet of

the Earth absolutely must be asserted as true; and the hypothesis that bestows to the Earth motion (either solely diurnal, or diurnal and annual) absolutely must be asserted as false and disagreeing with physical and indeed physico-mathematical demonstrations.[5]

In other words, in the age of the telescope, *science* shows the Copernican hypothesis to be wrong. And after this main conclusion, Riccioli includes copies of the condemnation of the Copernican system by church authorities in Rome, of the judgment against Galileo, and of Galileo's abjuration.[6]

In his analysis of the heliocentrism vs. geocentrism (of the hybrid sort) world-system debate in the *New Almagest*, Riccioli reviewed 126 arguments put forward by both sides. Edward Grant has described this as "probably the lengthiest, most penetrating, and authoritative analysis made by any author of the sixteenth and seventeenth centuries."[7] Forty-nine of the arguments Riccioli reviewed favored heliocentrism; seventy-seven favored geocentrism.[8] But to Riccioli, this was not merely a counting of arguments, for numbers of arguments were not what tipped the balance of Urania's scales.

The vast majority of arguments on both sides, he said, were unconvincing. Some were unconvincing because they were so bad (Riccioli says he is ashamed to bring up some, but does so because they have been used by various people).[9] Others were reasonable, but still unconvincing. For example, consider pro-heliocentrism argument number 22:

> The sun is the center of the Planetary System—as is demonstrated in the case of Mercury and Venus [by telescopic observations of their phases, illustrated in the frontispiece], and conjectured in the case of the others—so it ought to be the center of the Universe.[10]

This seems like a very reasonable argument. Riccioli notes, however, that the geocentrists have a valid answer to it. The answer is that the Sun is neither the center of the Moon's orbit, nor the center of motion for the fall of heavy bodies and the rise of fire, nor the center of the stars.[11] As another example, consider pro-geocentrism argument number 42, which says that the weighty Earth must be at the center of the universe because there is no explanation as to what would keep it in any other position. Here Riccioli notes that the heliocentrists can answer that the entire Earth has a natural circular motion about the center of the universe, in which its weight is not a factor.[12] Then there is pro-geocentrism argument number 53 (one of only two among the 126 that involve religious questions), which says that if Earth is not the center of the universe, then Hell is not at the lowest place, and someone going to Hell could conceivably ascend in doing so. How could someone *ascend* into Hell? Riccioli says that the answer to this argument is that Hell is a place defined by comparison to this world on which men[13] live and to God's Heaven; the relationship between Heaven, Hell, and the world of men is not affected by whether Earth moves.[14]

Riccioli did, however, find a select few arguments to be convincing— all of them anti-Copernican arguments for which the heliocentrists had no good answer. One of these concerned the question of detecting Earth's rotation. According to Riccioli, a rotating Earth must produce certain observable phenomena. These phenomena were, in fact, not observed. Thus the Earth must not rotate. These are the physical and "physico-mathematical" demonstrations Riccioli mentions in his main conclusion about which "hypothesis" can be asserted as being true.

A second anti-Copernican argument—one that Riccioli thought was stronger, even though it was technically answerable by the heliocentrists— concerned the sizes of stars. According to Riccioli, telescopic observations of the stars showed that, were the Copernican hypothesis correct, the stars would have to be huge. Indeed, he said, *one single star* in the Copernican world system could conceivably be larger than the *entire universe* in the hybrid geocentric world system. By contrast, in the hybrid geocentric system the stars would be reasonably sized. When the sizes of stars are taken into consideration, Copernicus's heliocentric world system looks absurd.

Riccioli devotes at least two chapters in the *New Almagest* to the star sizes issue, discusses the issue in a line of pro-geocentrism arguments (numbers 67–70),[15] and uses it to counter a pro-heliocentrism argument (number 9).[16] And while he grants to the heliocentrists the possibility that the above-mentioned demonstrations to detect Earth's rotation might not work because experiments could be insufficiently precise,[17] he grants no such escape to them on the problem of star sizes. Their only answer to this argument he says, is an appeal to Divine Omnipotence.[18] Technically, since he cannot deny the power of God, he cannot fully refute the Copernican answer on this matter. But, he says, "even if this falsehood cannot be refuted, nevertheless it cannot satisfy more prudent men."[19]

What is remarkable about Riccioli's analysis is that he was right. His analysis made sense, granted the knowledge available in his time. Today, in our time, you the reader have doubtlessly heard that Galileo used science to prove that the Earth circles the Sun—indeed, the back cover of the standard modern translation of Galileo's 1632 *Dialogue* explicitly states just that.[20] But I believe that by the time you finish the pages of this book, you will agree with Riccioli's *New Almagest* frontispiece. You will conclude that an objective and rational analysis of the best data on hand in the mid-seventeenth century would lead one to concur with Riccioli that the Copernican hypothesis, while certainly an improvement over the ideas of Ptolemy, did not compare favorably against hybrid geocentrism. You will find that some very strange ideas sprouted under Copernicanism—ideas about giant stars pointing to the nature of God, or indeed being the "warriors" of God. You will find that Riccioli's ideas can be traced back into the sixteenth century, to the great Danish astronomer Tycho Brahe, who promoted hybrid geocentrism. And you will find that between Brahe and Riccioli these ideas even played a role in the condemnation of the Copernican hypothesis by church authorities in Rome. Riccioli causes us to rethink our ideas of "heliocentrism vs. geocentrism" as being perhaps not so much about "science (and heliocentrism) vs. religion (and geocentrism)" as perhaps "science vs. science," or "religion vs. religion." Riccioli may even cause us to invert our view of the issue, wondering if at times the debate was actually a matter of "religion (and heliocentrism) vs. science (and geocentrism)." Giovanni Battista Riccioli invites a reexamination of what we believe we know about the Copernican Revolution.

2   The Universe that Riccioli Saw

To understand why Riccioli's ideas made sense, and why he invites a reexamination of what we think we know about the Copernican Revolution, we must understand the universe that he saw. Those of us who live in the twenty-first century's more developed nations are likely to know more about the Solar System than we know about the night sky. We probably know that Saturn has beautiful rings, or that Pluto used to be a called a planet before it was reclassified as a dwarf planet. We may know that these worlds have moons, and that Saturn is larger than Earth and that Pluto is smaller. We certainly know that Saturn and Pluto, along with Earth, circle the Sun. But relatively few of us can find Saturn among the stars of the night sky, using only our own two eyes. Relatively few know what constellation Saturn is in at the moment. And relatively few know that Saturn is visible to the eye, whereas Pluto is not, and thus we cannot find Pluto in the night sky using our own two eyes.

Many people today do not have a direct knowledge of the visible night sky. We are less aware of what the heavens look like than those who lived before electric lights were common—back in those days when, once the Sun went down, the night was truly dark and people saw the stars. Today the only people who might know the sky as well as people used to know it, before modern lighting brightened our nights, are amateur astronomers who spend a lot of time looking at the sky for its own sake, and perhaps those who live out in the country and farm for a living.

And even so, what amateur astronomers or farmers see may be influenced by what they have learned through modern science. If the farmer or the amateur does recognize Saturn in the night sky, and recognizes that it is among the stars of the constellation Taurus, for example, he or she probably thinks of it as "the world with the rings," and thinks of the stars of Taurus as unimaginably distant suns that we see as points of light of varying brightness. Yet looking at Saturn with the naked eye, there is nothing about it to indicate that it is a "world," and nothing to indicate that it has rings. There is nothing to indicate that the stars of Taurus are vastly more distant than Saturn, and nothing to indicate that they are suns. To understand Riccioli, we must learn to see the heavens as they appear on their own, without the filter of modern scientific knowledge.

What is the appearance of the heavens to the naked eye? First, consider that some things we see in the heavens are consistently there; some other things are not. Clouds, rainbows, shooting stars, and comets are not consistently there; these things are sometimes found in the heavens, and sometimes not.

Of those things that are found consistently in the heavens, the Sun and the Moon are the most prominent. They appear approximately the same size to the eye. This "apparent" size is about half of a degree, roughly the size a pea appears to have when held at arm's length. Between the point above your head (the "zenith") and the horizon there are 90 degrees. The Sun and Moon are the most prominent of heavenly bodies, but they are not large by comparison to the expanse of the heavens: 180 Moons could be lined up edge to edge between the horizon and the zenith.

The heavenly bodies that are most prominent numerically are the stars. Romantic songs may speak of millions of stars in the sky, but the stars visible to the naked eye actually number "only" in the thousands. The stars are fixed in position relative to one another. This is why we have constellations, such as Taurus (the bull) or Orion (the hunter) or Leo (the lion) or Scorpius (the scorpion, see fig. 2.1). These patterns in the stars were picked out by the ancient Greeks. They are still there today because the "fixed" stars have held their positions relative to one another for millennia.

As regards their apparent size, all the stars appear to the eye to be much smaller than the Moon and Sun. However, among the stars, the most prominent ones, such as Sirius (the Dog star), appear larger than the

Figure 2.1.   The stars of the constellation Scorpius, showing stars of various magnitudes. The most prominent star, just to the upper right of center, is the first magnitude star Antares. These figures were produced using the planetarium software Stellarium (http://www.stellarium.org), whose web site states that it "shows a realistic sky . . . just like what you see with the naked eye."

middling stars, such as Polaris (the North star), which in turn appear larger than the stars barely visible to the eye, such as Alcor (the "rider" of the "horse and rider" in the handle of the Big Dipper). The varying apparent sizes of stars is illustrated in figure 2.1.

Astronomers have not only traditionally named both the stars (Sirius, Polaris, Alcor) and the patterns they see in the stars (Taurus, Orion, Leo), but they have also classified the stars by their apparent sizes. The term traditionally used for the apparent size of a star is "magnitude" or "bigness." The most prominent stars were classified as being of the first magnitude—they were first-class stars. Lesser or second-class stars were of the second magnitude, and so forth down to the stars barely visible to the eye, which were of the sixth magnitude. The exact classification of stars into the various orders of magnitude was a matter of judgment, as the following old discussion shows:

> The *fixed Stars* appear to be of different Bignesses. . . . Hence arise the Distribution of *Stars*, according to their Order and Dignity, into *Classes;* the first Class . . . are called Stars of the first Magnitude; those that are next to them, are Stars of the second Magnitude . . . and so forth, 'till we come to the *Stars* of the sixth Magnitude, which comprehend the smallest *Stars* that can be discerned with the bare Eye. . . .

Altho' the Distinction of Stars into six Degrees of Magnitude is commonly received by *Astronomers;* yet we are not to judge, that every particular *Star* is exactly to be ranked according to a certain Bigness, which is one of the Six; but rather in reality there are almost as many Orders of *Stars,* as there are *Stars,* few of them being exactly of the same Bigness and Lustre. And even among those *Stars* which are reckoned of the brightest Class, there appears a Variety of Magnitude; for *Sirius* or *Arcturus* are each of them brighter than *Aldebaran* or the *Bull's* Eye, or even than the *Star* in *Spica;* and yet all these *Stars* are reckoned among the Stars of the first Order: And there are some *Stars* of such an intermedial Order, that the *Astronomers* have differed in classing of them; some putting the same *Stars* in one Class, others in another. For Example: The little *Dog* was by *Tycho* placed among the *Stars* of the second Magnitude, which *Ptolemy* reckoned among the *Stars* of the first Class: And therefore it is not truly either of the first or second Order, but ought to be ranked in a Place between both.[1]

The classification of stars using this "magnitude" scale is typically attributed to the ancient Greek astronomer Hipparchus of Nicea (190–120 B.C.). The idea of the apparent size or "bigness" of stars is clearly illustrated in this eighteenth-century description of the varying appearance of a new star, or "nova":

After [the nova] had thus absolutely disappeared, the place, where it had been seen, continued six months vacant. On the seventeenth of March following, the same observer saw it again, in exactly the same place, equal to a star of the fourth magnitude. On the third of April, 1671, the elder Cassini saw it, it was then of the bigness of a star of the third magnitude, he judged it to be a little less than [the star] in the back of the constellation; but, on the next day, repeating the observation, it appeared to him very nearly as large as that, and altogether as bright; on the ninth it was somewhat less; on the twelfth it was yet smaller, it was then less than the two stars at the bottom of Lyra; but, on the fifteenth, it had increased again in bigness, and was equal to those stars; from the sixteenth to the twenty-seventh of the same month he observed it with a peculiar attention; during that period it

changed bigness several times, it was sometimes larger than the biggest of those two stars, sometimes smaller than the least of them, and
sometimes of a middle size between them. On the twenty-eighth of
the same month it was become as large as the star in the beak of the
Swan, and it appeared larger from the thirtieth of April to the sixth of
May. On the fifteenth it was grown smaller; on the sixteenth it was of a
middle size between the two, and from this time it continually diminished till the seventeenth of August, when it was scarce visible to the
naked eye.[2]

We have now discussed the general appearance, to the naked eye, of
the Sun, Moon, and fixed stars. A person who observes the sky for even a
few hours at night will also notice these bodies all share a basic daily or "diurnal" motion—they rise in the east and set in the west. Moreover, to the
casual observer the stars, Sun, and Moon appear to all move as a unit.
However, if one of the stars is observed to rise at, for example, 11:00 p.m.
on a given night, on the following night it will rise at 10:56 p.m.—the time
from starrise to starrise is 23 hours, 56 minutes. By contrast, timing sunrise
to sunrise yields 24 hours—one day. Timing moonrise to moonrise yields
24 hours, 48 minutes. (These values hold true near the equator, with a flat
horizon, such as at the ocean. The picture gets more complicated at locations progressively farther from the equator, and with a progressively more
obstructed horizon. However, these values still hold in the average.) In
short, we see the Sun, Moon, and stars circling around us, with the stars
moving the most rapidly and the Moon the most slowly. The result is that
the Sun and Moon drift steadily through the stars—and in particular
through the constellations of Leo, Cancer, Virgo, Sagittarius, and so forth,
that comprise the constellations of the zodiac. The Sun, drifting 4 minutes
against the stars each day, takes a year to cycle fully through the zodiac; the
Moon, drifting 52 minutes (48 minutes vs. the Sun, plus the Sun's 4 minutes vs. the stars) against the stars each day, takes a month.

   Everything described so far can be observed by any interested person
who dedicates a couple of weeks to careful skywatching. What requires
more time to be observed is the constancy of the heavens. The stars do rise
and set and rise again, but they don't move around with respect to each
other. The fixed stars pretty much always look the same. To the naked eye,

stars are not created and not destroyed: we know of no stars visible to the naked eye today that were not visible in ancient times, and we know of none visible in ancient times that are not still visible today (this often surprises students, who have heard of star birth and death through planetarium visits and educational videos, but the life span of stars is so long that those visible to the eye have not noticeably changed over centuries). The fixed stars are not the only things in the heavens that remain constant. The cycles and motions of the heavens also do not change. Two centuries ago the Sun did not rise in the south and set in the north; it rose in the east and set in the west, just like it does today. The days were not 28 hours long in the time of Christopher Columbus; they were 24 hours long, just like today. The Moon did not require half a year to pass through the zodiac when the pharaohs ruled Egypt; it required a month, just as it does today.

The constancy of the cycles of the heavens is something that is part of our culture and language. If we want to say that something is certain, we will refer to it being "sure as the Sun will rise." Indeed, the idea of being as certain as the workings of the heavens even shows up in advertising and pop culture: Days Inn offers "A Promise as Sure as the Sun," and the theme song from the Disney movie *Beauty and the Beast* makes references to being "certain as the Sun rising in the East." There is nothing—*nothing*—in human experience as constant as the heavens. Old trees die and fall, and new ones grow up. A stand of native wildflowers that seems like it has been here for ages is overrun by an invasive species. Structures and monuments raised with the greatest of skill deteriorate and decay. The most finely engineered machines eventually break down and require repair. Forests that have stood for centuries are cut down and made into farmland. Land that has been farmed for generations is abandoned and returns to forest. Rivers that have flowed for ages untold are dammed up into lakes. Mountains that seem as old as the Earth itself are blasted through for highways. Change is part of life on Earth, as Percy Bysshe Shelley captured in his poem "Ozymandias":

> I met a traveler from an antique land,
> Who said: Two vast and trunkless legs of stone
> Stand in the desert. . . . Near them, on the sand,
> Half sunk a shattered visage lies, whose frown,

And wrinkled lip, and sneer of cold command,
Tell that its sculptor well those passions read
Which yet survive, stamped on these lifeless things,
The hand that mocked them, and the heart that fed:
And on the pedestal, these words appear:
"My name is Ozymandias, King of Kings:
Look on my Works, ye Mighty, and despair!"
Nothing beside remains. Round the decay
Of that colossal wreck, boundless and bare
The lone and level sands stretch far away.[3]

But even as all on Earth changes, the heavens remain constant, like nothing else we ever see. This is not something that can be observed in a short span of time, but it is something humanity has observed for itself over the ages. The Sun rose in the east and set in the west on Ozymandias, too. His days were 24 hours long, like ours.

So how do we explain all this? We see that the Sun rises in the east and sets in the west. But why does it rise and set? What makes it move? For Riccioli, such things were explained through Aristotelian ideas.

According to Aristotelian thinking, the universe is made of five substances, or elements, whose properties explain all natural motion. These elements are earth, water, air, fire, and the quintessence or "fifth element." Elemental earth has the property of "gravity" or "heaviness"—it has a natural tendency to move toward the center of the universe (the center of the universe being the center of the circular motions of the Sun, Moon, and stars). Identically sized spheres made of materials of differing densities—lead, iron, stone (more or less concentrated matter)—have differing amounts of gravity. Elemental water also has gravity—but less than elemental earth. Elemental fire possesses "levity"—it has a natural tendency to move away from the center of the universe, at least to some extent. Air has neither the gravity of water nor the levity of fire.

Anyone interested in seeing all this in action can do so easily enough. Fill a bottle one third with sand and one third with water and leave one third filled with air. Stop up the bottle and shake it, and then let it stand. The sand will settle out to the bottom, with water above the sand, and air above the water. If one could somehow keep a fire burning in the bottle while doing this, the products of the fire would rise to the very top of the bottle, above the air.

The natural motions of these four elements are all straight. If not moved by force in some other way, elemental fire moves straight up away from the center of the universe, and elemental earth straight down toward it. Thus in still air a candle flame rises straight up, and a rock falls straight down. It is important to note that Aristotelian gravity is not a mutual attraction between rocks, or between the Earth and the rock. Rather, it is a natural property of the elemental earth that makes up the rock to move toward the center of the universe—toward the lowest point in the universe. A rock would move toward the center of the universe whether the Earth was there or not. In fact, the Earth lies at the center of the universe because that is the place where all the heavy stuff in the universe accumulates. All the heavy material is piled up in a dirty ball at the center of the universe, and that is the Earth.

Of course materials can be forced to move against their natural motions. A rock's natural tendency may be to fall straight down to the ground—that is, to move toward the center of the universe, stopping when it encounters the other earthly material that is already piled up around the center—but the rock can be carried, or pushed, or thrown, so that it moves contrary to its natural motion. Aristotle had argued that such motion always requires a mover to make it go, but in Riccioli's time the concept of "impetus" was in broad use.

The idea of impetus had been developed by Jean Buridan in the first half of the fourteenth century. A challenge for Aristotelian thinking was the motion of an object that continues moving once set in motion, such as a stone that has been tossed. Once the stone leaves the hand of the person tossing it, should it not fall straight to the ground, because it no longer has a hand moving it? Why does it continue its unnatural motion? Aristotle explained the tossed stone's flight by saying that the air through which the stone flies becomes the mover of the stone. Air carries the stone along its trajectory as the stone moves through the air. If objects move only if powered by a mover, Aristotle asks,

> how is it that some of them, like things thrown, are continuously in motion when the mover is not touching them? . . .

We must then say this, that the first mover causes air or water or some other such object, which by nature can move another and be moved by another, to be like a mover. But this object does not simultaneously cease being a mover and an object moved. It ceases simultaneously being moved when the mover ceases causing it to be moved, but it may still be a mover; and in view of this, it may cause some other consecutive object to be moved (and the same may be said of this other object).[4]

Aristotle called this process "circular displacement," and said it occurred in air and water.[5]

Buridan used the motion of a round spinning toy top or a freely spinning grinding wheel to argue against Aristotle. Buridan argued that air cannot be what keeps round spinning objects in motion, because such objects spin in place. Air is not displaced by their motion, like it is in the case of a tossed stone. Buridan even experimented with trying to block air from flowing around a spinning object in any way, in order to see if that caused the object to stop spinning. It did not. Buridan further argued that if air was what carried a tossed object through its trajectory, then it would be possible to throw a feather farther than a stone, which it is not.[6]

So Buridan put forth a theory of motion that was contrary to the Aristotelian idea that anything that moves requires a mover. He argued that air simply resists motion, and that the reason a tossed stone flies through the air is because the hand, in throwing the stone, imparts to the stone a kind of self-moving action. Buridan called this action "impetus." The impetus of the stone keeps it moving despite the resistance of the air. Buridan described impetus as being in the direction the stone is moving, as being greater (all else being equal) in a heavier stone than in a lighter stone moving at the same speed, and as being greater (all else being equal) in a faster-moving stone than in a slower-moving stone of the same weight. Thus, said Buridan, when you throw a stone, you give it impetus, and the stone keeps moving owing to that impetus. The stone's impetus is continually decreased by the resistive action of the air, while the stone's gravity turns the direction of the impetus downward, so that eventually the stone drops to the ground. Buridan's discussion of this is most informative:

Thus we can and ought to say that in the stone or other projectile there is impressed something which is the motive force of that projectile. And this is evidently better than falling back on the statement that the air continues to move the projectile. For the air appears rather to resist. Therefore, it seems to me that it ought to be said that the [mover] in moving a moving body impresses in it a certain impetus or a certain motive force of the moving body, in the direction toward which the mover was moving the moving body, either up or down, or laterally, or circularly. And by the amount the [mover] moves that moving body more swiftly, by the same amount it will impress in it a stronger impetus. It is by that impetus that the stone is moved after the projector ceases to move. But that impetus is continually decreased by the resisting air and by the gravity of the stone, which inclines it in a direction contrary to that in which the impetus was naturally predisposed to move it. Thus the movement of the stone continually becomes slower, and finally that impetus is so diminished or corrupted that the gravity of the stone wins out over it and moves the stone down to its natural place.

This method, it appears to me, ought to be supported because the other methods do not appear to be true and also because all the appearances are in harmony with this method.

For if anyone seeks why I project a stone farther than a feather, and iron or lead fitted to my hand farther than just as much wood, I answer that the cause of this is that the reception of all forms and natural dispositions is in matter and by reason of matter. Hence by the amount more there is of matter, by that amount can the body receive more of that impetus and more intensely. Now in a dense and heavy body, other things being equal, there is more of prime matter than in a rare and light one. Hence a dense and heavy body receives more of that impetus and more intensely, just as iron can receive more calidity [heat] than wood or water of the same quantity. Moreover, a feather receives such an impetus so weakly that such an impetus is immediately destroyed by the resisting air. And also if light wood and heavy iron of the same volume and of the same shape are moved equally fast by a projector, the iron will be moved farther because there is impressed in it a more intense impetus, which is not so quickly corrupted as the lesser impetus would be corrupted. This also is the reason why it is more difficult to bring to rest a large smith's mill [like a grinding wheel] which is moving swiftly than a small one, evidently because in the large one, other things being equal, there is more impetus. And for this reason you could throw a stone of one-half or one pound weight farther than you could a thousandth part of it. For the impetus in that thousandth part is so small that it is overcome immediately by the resisting air.[7]

In Riccioli's time, three centuries after Buridan, impetus was a term that appeared often in discussions of physics. However, it appeared in a form less crisply defined than Buridan's clear definition, in which impetus is a specific quantity that depends proportionately on weight and speed, and that has direction. But despite this corruption, the general physics of impetus was available to explain forced or nonnatural motion: set something in motion, and it will keep moving, owing to its weight and speed, at least for a while.

All this discussion has been in regards to the motion of things on Earth—things composed of earth, water, air, and fire. As regards the heavens, the realm of the fifth element, the Aristotelian idea was that natural motion is circular and unending. Since the fifth element is found only in the heavens, it can be as light, as incorruptible, as strong, as brilliant, as easy to move, as might be needed to explain the heavenly motions. That the heavens are made of a different substance than Earth is obvious— here on Earth everything changes, all lights burn out, and every wagon and every machine eventually stops moving. By contrast, the heavens never change, ever. The celestial lights never go out. The celestial motions never cease. Aristotle postulated a Prime Mover to drive the celestial machinery, a Prime Mover that he linked to the action of an eternal, unchanging, perfect God. Buridan, by contrast, thought the same concept that explained the motion of tops and stones could be used to explain the motions of the heavens, especially if the heavens contained no air or other resisting substance. Buridan wrote that the Bible does not state that the heavens are moved by a mover, so it could well be that

God, when He created the world, moved each of the celestial orbs as he pleased, and in moving them He impressed in them impetuses which moved them without his having to move them any more except by the method of general influence whereby he concurs as a co-agent in all things which take place; "for thus on the seventh day He rested from all work which He had executed by committing to others the actions and the passions in turn." And these impetuses which He impressed in the celestial bodies were not decreased nor corrupted afterwards, because there was no inclination of the celestial bodies for other movements. Nor was there resistance which would be corruptive or repressive of that impetus.[8]

Thus the ideas of Aristotle, mixed with some ideas about impetus, explained the *behavior* of the universe. But what about the *structure* of the universe? How would Riccioli have understood that? A few basic points about the structure of the universe were evident.

First, the Earth must be round—the only logical shape expected for something composed of material that is trying to crowd from all directions into the lowest point in the universe. This is easily verified by observing the round shadow of Earth passing across the Moon during an eclipse, as seen in figure 2.2. Moreover, it is verified by the fact that, as one heads north, Polaris (the North star or pole star) rises toward the zenith in the northern sky, and as one heads south, it sinks toward the northern horizon. Indeed, by noting that each 70 miles of northward or southward travel results in one degree of change in the height of Polaris, it is possible to determine the size of the Earth: each 70 miles is 1/360 of Earth's circumference, so Earth measures 70 miles x 360 = 25,200 miles around. This large world consists of elemental earth, with water sitting on it, and air above these. Elemental fire rises up toward the Moon. The heavens— the Moon and everything above the Moon—are composed of the amazing fifth element.

Second, the heavens must be huge. If the Earth was of any size compared to the heavens, then we would see less than half the sky at any given time. But we can map the stars, and we see that, at least where the horizon is clear, half the sky is always visible. For this to be possible, the Earth must be so small as to be almost a point by comparison to the sphere of the stars that encircles it (see fig. 2.3).

Third, the motions of the heavens are greatest for those celestial bodies farthest from Earth, and least for those nearest the heavy, immobile

Figure 2.2. The Moon partially eclipsed by Earth's shadow on 21 December 2010, seen with the Washington Monument in the foreground. The curve of the shadow of the Earth indicates that Earth is round. Photo credit: NASA/ Bill Ingalls. Available at http://www.flickr.com/photos /nasahqphoto/5279396149/ in/set-72157625518083075.

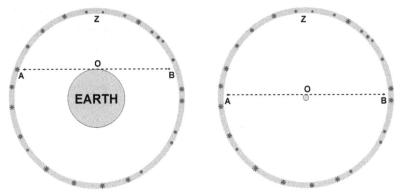

Figure 2.3. Left: Were the Earth of significant size compared to the distance to the stars, then an observer at O would see only the span of the starry heavens from A on one horizon, through the zenith Z, to B on the opposite horizon—well less than half the celestial sphere could be seen at any given moment.

Right: That half the heavens are seen at any given moment demonstrates that the Earth is in fact not of significant size compared to the distance to the stars. Here the span of the starry heavens A through Z to B as seen by the observer at O is nearly half the celestial sphere, and were the Earth so small as to be nearly a point in this figure, the AZB span would be exactly half the celestial sphere.

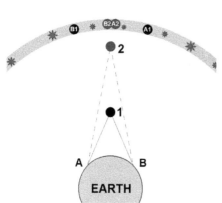

Figure 2.4.
Object 1 appears among the stars at A1 as seen by an observer at A, but it appears at B1 as seen by an observer at B. The difference between the two positions is the *diurnal parallax* of the object. Object 2 is more distant from Earth and thus has a smaller diurnal parallax—there is a smaller difference in the positions of object 2 as seen by the two observers.

Earth. The stars move the fastest (circling Earth once every 23 hours, 56 minutes), followed by the Sun (24 hours), and the Moon (24 hours, 48 minutes). The distances of the Sun and Moon are known through methods such as "diurnal parallax"—the difference in the position of a celestial body as measured by observers at two locations on Earth (fig. 2.4). The Earth sits at rest in the center of them all.

This is the universe as Riccioli would discuss it, in terms that Riccioli would understand. From here forward, we will attempt to inhabit it. Yes, today science tells us that gravitation is a universal mutual attraction among matter, an attraction that grows proportionately with mass and weakens proportionately with the square of distance, and that it is *not* a tendency of elemental earth to move to the center of the universe. Today science tells us that, were the Earth not here, a rock would *not* fall. Today we understand that "an object in motion remains in motion" unless acted on by a net external force. Today astronomers still use the term "magnitude" in describing stars, but that term has been redefined as a measure of the intensity of light: astronomers still say Polaris is magnitude 2, and Alcor magnitude 4, but astronomers rate Sirius as magnitude −1.5, and the Sun as −26.7 (and none of these are matters of judgment, but of precise measurement). Moreover, astronomers do not speak of the apparent sizes of stars at all, saying rather that they are points of light of differing brightness. We do this because that is what we know them to be; we know that the apparent sizes of the stars are an illusion caused by the physics of light. Today we understand that heavenly bodies such as the Sun and the Moon are made of the same elements that comprise the Earth, although we do still posit heavenly substances to explain what we see in the universe ("dark matter" and "dark energy" have more than a little in common with "the fifth element").

But all of these things were discovered after Riccioli. Universal gravitation and "an object in motion remains in motion" are products of Newtonian physics (although Buridan might lay claim to the latter).[9] Isaac Newton would not develop them until after Riccioli's death. The modern view of stars as points of light, and the redefining of the magnitude system as a measure of the intensity of those points, would not come into existence until the nineteenth century. Understanding Riccioli requires putting aside these and all the other discoveries that would be made after his time, and seeing the universe in the Aristotelian-plus-impetus terms in which he, and so many others, saw it.

3   The Anti-Copernican Astronomer

The stars, to the naked eye, present diameters varying from a quarter of a minute of space, or less, to as much as two minutes. The telescope was not then invented which shows that this is an optical delusion, and that they are points of immeasurably small diameter. It was certain to Tycho Brahé, that if the earth did move, the whole motion of the earth in its orbit did not alter the place of the stars by two minutes, and that consequently they must be so distant, that to have two minutes of apparent diameter, they must be spheres as great a radius at least as the distance from the sun to the earth. This latter distance Tycho Brahé supposed to be 1150 times the semi-diameter of the earth, and the sun about 180 times as great [by volume] as the earth. Both suppositions are grossly incorrect; but they were common ground, being nearly those of Ptolemy and Copernicus. It followed then, for any thing a real Copernican could show to the contrary, that some of the fixed stars must be 1520 millions of times as great as the earth, or nine millions of times as great as they supposed the sun to be. . . . Delambre, who comments with brief contempt upon the several arguments of Tycho Brahé, has here only to say, "We should now answer that no star has an apparent diameter of a second." Undoubtedly, but what would you have answered then, is the reply. The stars were spheres of visible magnitude, and are so still; nobody can deny it who looks at the heavens without a telescope; did Tycho reason wrong because he did not know a fact which could only be known by an instrument invented after his death?

—from the article "Brahé, Tycho" in the 1836 *Penny Cyclopædia
of the Society for the Diffusion of Useful Knowledge*

Riccioli's anti-Copernican arguments seem to have their roots in the work of the great Danish astronomer Tycho Brahe (1546–1601). Tycho, a nobleman of the generation between Copernicus and Galileo, was widely recognized as the finest astronomer of his era. While other well-known astronomers like Copernicus or Galileo made their observations and published their results and ideas as individuals (Copernicus did astronomy on the side while working as a canon at the Frauenburg cathedral, for example), Tycho ran a major observatory (Uraniborg) and research program on his island of Hven. Different writers have described the cost of these to the Danish crown as being proportionately comparable to the budget of NASA.[1] Tycho's observatory was the best of its day—a Hubble Space Telescope of his time (fig. 3.1). His observational data were unsurpassed in quality and quantity. It was Tycho's data that would enable Johannes Kepler to work out his famous laws of planetary motion, and thus bring about our modern understanding of the solar system. When Albert Curtius published his 1666 *Historia Coelestis*, a 977-page catalog of observed planetary positions over time, from antiquity to 1630, Tycho's data occupied the overwhelming majority of the book, with the combined data gathered by all other astronomers over the centuries occupying the remaining pages.[2] With access to the biggest and best available instruments, and with the most skilled assistants, Tycho could achieve incredible accuracy in his work.

Tycho worked before the advent of the telescope. His instruments were nontelescopic, or "naked eye" instruments (see an example of one in fig. 3.2). Unlike a telescope, they did not enhance what an observer saw in the heavens. They simply allowed an astronomer to precisely measure and record what he or she was seeing (the use of "she" here is not merely a nod to inclusive language; the late seventeenth-century astronomer Elisabeth Hevelius used such instruments; see fig. 3.3). A ruler is an instrument like that. A ruler does not change what the eye sees in any way. It does not improve the eye's vision. It does not enhance the eye's ability to see detail. A ruler simply allows its user to measure and record what the eye of the user sees. One needs no ruler to determine that a pencil is longer than an acorn, or even that a pencil is several times longer than an acorn—eyes alone are sufficient for that. But a ruler does allow one to determine these sizes to some precision, even though it does not at all change what the eyes see. A truly fine ruler, perhaps outfitted with a sliding set of jaws and a vernier scale on it so that it is now a sort of "slide caliper" (fig. 3.4), will allow one to determine the sizes of the pencil and acorn with impressive

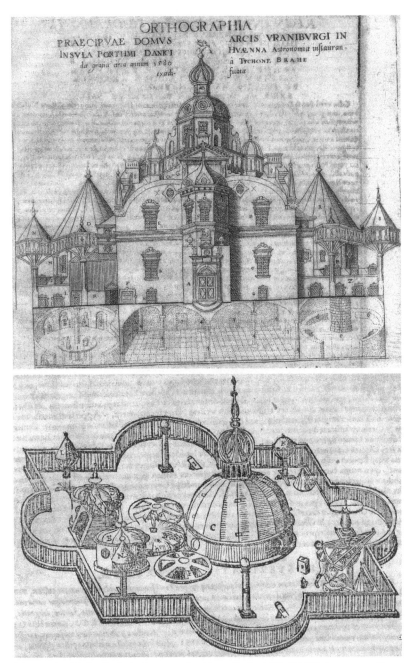

Figure 3.1. Tycho Brahe's observatories (from Brahe 1602, 97, 99). Images credit:
ETH-Bibliothek Zürich, Alte und Seltene Drucke.

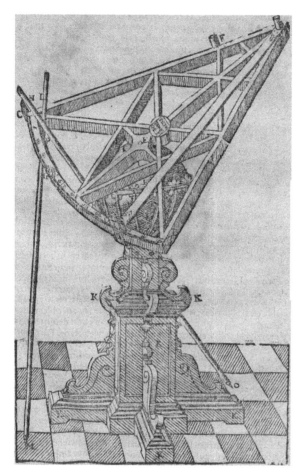

Figure 3.2.
One of Tycho
Brahe's nontelescopic
instruments (from
Brahe 1602, 53).
His instruments
were designed and
used mostly for the
purpose of making
precise measurements
of positions,
separations, and
diameters of celestial
bodies. Image credit:
ETH-Bibliothek
Zürich, Alte und
Seltene Drucke.

precision. Tycho Brahe's instruments were like the slide caliper ruler—
they allowed the user to make impressively precise measurements of what
the eye saw, without changing what the eye saw.

Modern analysis of Tycho's work illustrates this precision. Tycho could
define any star's position within a circle whose diameter was less than a
minute of arc (1/60 degree, or 1/30 the apparent diameter of the Moon).
For certain sorts of measurements, such as the altitude of the North Celestial Pole, he could exceed this accuracy by better than a factor of ten.[3] Owen
Gingerich, who often illustrates Tycho's incomparable contribution to the
astronomy of his time by means of the *Historia Coelestis*, has argued that

Figure 3.3. Astronomers Johannes and Elisabeth Hevelius using a nontelescopic measuring instrument (Hevelius 1673, after 222). Image credit: ETH-Bibliothek Zürich, Alte und Seltene Drucke.

Figure 3.4. A vernier caliper—a device that allows for accurate measurement of the sizes of objects but that does not magnify or otherwise alter what the eye sees. Image credit: M. Colcher (Wikimedia Commons). Available at http://commons.wikimedia.org /wiki/File:Pied_Coulisse.png.

Tycho's quest for better observational accuracy "places him far more securely in the mainstream of modern astronomy than Copernicus himself."[4] Gingerich notes that "only twice in the history of astronomy has there been such an enormous flood of new data that just changed the scenes"—the flood from Tycho Brahe and the flood from today's digital revolution.[5]

Tycho saw in the Copernican world system an idea that had merit, but that also had significant flaws. Copernicus had proposed an explanation for the complex motions of the five "wandering stars"—Mercury, Venus, Mars, Jupiter, and Saturn (the word "planet" comes from the Greek word for "wander"). These did not remain in any one constellation, but instead drifted through the zodiac constellations with looping, complex motions, while their magnitudes varied as though they were moving closer to and farther from the Earth. Copernicus said all this could be better explained if those bodies, and Earth, circled the Sun, rather than if those bodies, and the Sun, circled Earth. But Copernicus proposed this hypothesis a century and a half before Isaac Newton, in a world in which physics followed the ideas of Aristotle (as described in chapter 2). The Copernican system's moving Earth was contrary to accepted physics. There was simply no mechanism by which the Earth, a vast body made of heavy materials, could be easily moved. How could the Earth—countless cubic miles of very heavy rock—be moved, and kept in motion, when it was so difficult just to move a loaded wagon down a street? The motions of heavenly bodies, by contrast, were easy to explain—they were made of the aethereal, light, strong fifth element. The properties of the fifth element were such that heavenly bodies moved naturally. Thus Tycho said,

> [The Copernican system] expertly and completely circumvents all that is superfluous or discordant in the system of Ptolemy. On no point does it offend the principles of mathematics. Yet it ascribes to the earth, that hulking, lazy body, unfit for motion, a motion as quick as that of the aethereal torches, and a triple motion at that.[6]

He also noted that motion "belongs to the sky itself whose form and subtle and constant matter are better suited to perpetual motion, however fast," and not to the Earth, a body that is very heavy and dense.[7]

Tycho produced his own hypothesis regarding the structure of the universe, one which he may have felt retained the Copernican system's lack of Ptolemaic superfluity or discordance, but which avoided ascribing motion to a hulking, lazy, heavy, dense Earth. In Tycho's hypothesis the Sun, Moon, and stars circled an immobile Earth, while the planets circled the Sun (fig. 3.5). This is the hybrid geocentric world system seen in Riccioli's frontispiece in chapter 1 (figs. 1.1 and 1.3).[8]

Today the hybrid geocentric hypothesis is often referred to as the "Tychonic" hypothesis, but hybrid geocentrism was an old idea. Venus and Mercury never wander far from the Sun in the sky, and since the days of Aristotle various people had proposed that this is because they actually

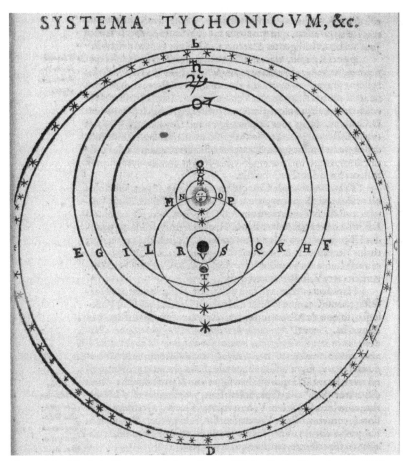

Figure 3.5.   The hybrid geocentric hypothesis of Tycho Brahe (from Locher 1614, 52). Mercury, Venus, Mars, Jupiter, and Saturn circle the Sun as in the Copernican hypothesis, while the Sun circles the Earth (as do the Moon and stars). Image credit: ETH-Bibliothek Zürich, Alte und Seltene Drucke.

circle the Sun. The idea seems to have originated at the time of Aristotle, and was passed down in one way or another through various writers over time. One of these, Martianus Capella, was mentioned by Copernicus in his *On the Revolutions of Heavenly Spheres*, and so this idea is often called the Capellan hypothesis.[9] Tycho expanded on this idea to have all of the wandering stars circling the sun. The Tychonic hypothesis was mathematically identical to that of Copernicus, insofar as the Sun, Moon, and planets were concerned. Therefore, like the Copernican system, it would be compatible with Galileo's telescopic discoveries (Venus circled the Sun in both, for example). In addition, it did not conflict with scriptural passages that spoke of an immobile Earth or a moving Sun. Thus, Brahe said, it was an idea that "offended neither the principles of physics nor Holy Scripture."[10]

Yet Tycho's opposition to the Copernican system was not merely a matter of adherence to scripture or to Aristotelian ideas. Brahe also produced at least one very robust anti-Copernican argument, and produced at least the seeds of another. These would form the basis for Riccioli's arguments decades later.

Brahe's most potent argument was based on the magnitude or "bigness" of the stars, discussed in chapter 2. In the stars, Brahe discovered a very big problem with the Copernican system. If, like Copernicus said, the Earth moves annually about the Sun, and the Sun and the stars do not move, then Earth's annual motion should reveal itself in the stars—a phenomenon known as *annual parallax*. As orbital motion carries Earth toward certain stars and away from others, those stars should grow and decrease in magnitude, respectively. Their relative positions should change, too, as Earth draws nearer and moves farther away—two neighboring stars should appear at least somewhat more widely separated when Earth is nearer to them than when it is farther from them (fig. 3.6). Indeed, Copernicus said that the variations in brightness and the looping motions of the planets (wandering stars) were manifestations of just such effects. And yet, said Copernicus,

> None of these phenomena appears in the fixed stars. This proves their immense height, which makes even the sphere of the annual motion, or its reflection, vanish from before our eyes. For, every visible object has some measure of distance beyond which it is no longer seen, as is demonstrated in optics. From Saturn, the highest of the planets, to the sphere of the fixed stars there is an additional gap of the largest size.[11]

Yet more than just the Saturn-to-stars gap was of "the largest size" in the Copernican system. Here was where Tycho built his powerful argument against Copernicus.

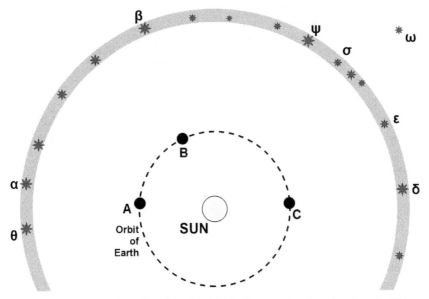

Figure 3.6. Annual parallax. If Earth orbits the Sun, moving from A to B to C within a six month period, then a number of observable changes should be seen in the appearance of the stars. When Earth is at A, star α will be closer and thus should appear larger than star β; two months later, when Earth is at B, the situation will be reversed. In short, the apparent size (magnitude) of a star should vary as the months pass. In fact, this does not happen. Likewise, when Earth is at B, stars θ and ψ will be opposite each other in the sky—one rising as the other sets. Thus the visible expanse of the heavens will stretch only from θ through α and β to ψ—less than half the expanse of starry heaven. In fact, half the stars are visible at any given time. As the Earth moves under the stars, its pole should not always point to the same place among the stars, and the overall positions of the stars as seen from Earth should be altered. In fact, the Earth's pole points consistently toward Polaris over the course of a year, and the overall positions of stars do not change over the course of a year. If the stars extend into space, then stars σ and ω will appear close to one another in the sky (a "double star") when the Earth is at B, but less so when Earth is at C. In fact, no such yearly changes occur in double stars. The Copernican answer to why these changes are not seen is that the orbit of Earth is of negligible size compared to the distance to the stars: thus all these annual parallax effects exist, but are negligibly small.

	Apparent Diameter		Distance		Physical Radius	Physical Volume
	min	*sec*	*E. R.*	*S. D.*	*E. R.*	*E. V.*
Moon	33		60	0.05	0.29	0.02
Sun	31		1150	1	5.19	139.4
Mercury	2	10	1150	1	0.36	0.05
Venus	3	15	1150	1	0.54	0.16
Mars	1	40	1745	1.52	0.42	0.08
Jupiter	2	45	5990	5.21	2.4	13.75
Saturn	1	50	10550	9.17	2.81	22.26
1st mag	2		14400	12.52	4.19	73.5
2nd mag	1	30	14400	12.52	3.14	31.01
3rd mag	1	5	14400	12.52	2.27	11.68
4th mag		45	14400	12.52	1.57	3.88
5th mag		30	14400	12.52	1.05	1.15
6th mag		20	14400	12.52	0.7	0.34

Table 3.1.   Tycho Brahe's apparent sizes of and average distances to celestial bodies, with resulting physical sizes *assuming a hybrid geocentric cosmos* (E. R. = Earth Radius; S. D. = Solar Distance; E.V. = Earth Volume). Apparent diameter and distance values are from Brahe 1602, 424–31; also see Dreyer 1890, 190–91, and Thoren 1990, 302–7. Physical radius and volume are calculated from these, and so may differ slightly from Tycho's numbers, which suffer from rounding and typographical errors.

Tycho obtained precise measurements of the apparent diameters of the different celestial bodies, and also determined their distances from Earth. His results for a geocentric world system are summarized in table 3.1 and illustrated in figure 3.7. As seen in the table, Tycho measured a first magnitude star, for example, to have an apparent diameter of 2 arc minutes, or approximately 1/15 the diameter of the Moon. This value was not all that different from the values measured by Ptolemy and others.[12] The planet data in the table are for representative planetary distances that are the average of each planet's two extremes from Earth. For Mercury and Venus, this is simply 1 Solar Distance (S. D.). For Mars, Jupiter, and Saturn, this equals the radius of their circles of motion around the Sun. Planetary distances in S. D. could be worked out by observing planetary motions—

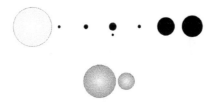

Figure 3.7.   The relative sizes of celestial bodies calculated by Tycho Brahe, based on his observations and measurements, for (from left to right, top row) the Sun, Mercury, Venus, Earth and Moon, Mars, Jupiter, Saturn, as well as for (bottom row) a large star and a mid-sized star in a hybrid geocentric universe (where the stars lie just beyond Saturn, as in fig. 3.5). Sun, stars, and planets all fall into a fairly consistent range of sizes. From Graney 2013a.

measuring the maximum angle between Mercury and the Sun yields the radius of Mercury's circle, for example—and the values given in the table generally agree with modern values. However, relating the solar distance to the radius of the Earth (E. R.) was problematic. All methods of determining this value that might work in theory (such as determining the lunar distance in E. R. via diurnal parallax, and then exactly measuring the angle between the Moon and Sun when the Moon is precisely at first quarter phase, and then determining the ratio of solar to lunar distances based on how much less that angle is than 90 degrees) were highly prone to error in practice. Tycho used a value of 1150 E. R. = 1 S. D., which is too small by more than a factor of twenty, and which leads to a physical size for the Sun that is likewise too small. However, Tycho's value was in line with the values used by astronomers from Ptolemy in ancient times to Copernicus; Galileo would use a similar number as well.[13]

By contrast, Tycho's stellar distances for a geocentric world system are not based on measurement at all. In a geocentric cosmos there is virtually no means of determining stellar distances from Earth. Tycho assumed the fixed stars lay beyond the furthest retreat of Saturn, but that could be 11 S. D. as easily as his 12.52 S. D. seen in table 3.1[14]—and were they at 11 S. D. instead of 12.52 S. D. the physical sizes of the stars would be slightly smaller (fig. 3.8). Thus the physical star sizes Tycho gives are necessarily estimates, based on an assumed interval between Saturn and the fixed stars. The end result of Tycho's measurements and calculations was

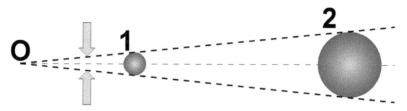

Figure 3.8.   The relationship between the apparent size, the physical or true size, and the distance of a celestial body. Tycho Brahe measured stars to have a certain apparent size (indicated by arrows). However, the farther a given star of a certain apparent size is from an observer (O) on Earth, the physically larger it must be: if the star is farther from Earth (at location 2), it will have a larger true size than if it is closer to Earth (at location 1).

that, in a geocentric cosmos, the sizes of bodies ranged from the Sun on one end down to the Moon on the other (again, see figure 3.7). The fixed stars fell nicely into that range: first magnitude stars were just a little smaller than the Sun.

But in a heliocentric cosmos, the distance to a star *can* be determined geometrically, by means of measuring the star's annual parallax. The farther away a star is, the smaller its parallax will be. And even if annual parallax is not observed, its nondetection can be used to establish a minimum stellar distance. In order for annual parallax to be no more than a minute of arc (just falling under Tycho's circle of general accuracy, and thus just evading detection), the distance to the fixed stars would have to be almost 7,000 S. D. Copernicus's Saturn-to-stars "gap of the largest size" would be over 700 times the Sun-to-Saturn distance. And the stars themselves, rather than falling within the size range of the other heavenly bodies, would have to be hundreds of times the diameter of the Sun (table 3.2, fig. 3.9). Why? Because the farther away an object of a given *apparent* size is, the larger it must be in terms of actual *physical* size. The Sun and Moon have roughly the same apparent size, as can be seen on an evening when the Sun is setting while the Moon is rising. But as the Sun is much farther away than the Moon, it is actually the much larger body. By the same geometry that applies to the Sun and Moon, the farther away the stars were, the larger they had to be physically (fig. 3.8). Were the stars 7,000 S. D. distant, they would have to be enormous—an ordinary third-magnitude

	Apparent Diameter		Distance		Physical Radius		Physical Volume
	min	sec	E. R.	S. D.	E. R.	S. D.	E. V.
1st mag	2		7906818	6875	2300	2	12167000000
2nd mag	1	30	7906818	6875	1725	1.5	5132953125
3rd mag	1	5	7906818	6875	1246	1.08	1933658782
4th mag		45	7906818	6875	863	0.75	641619141
5th mag		30	7906818	6875	575	0.5	190109375
6th mag		20	7906818	6875	383	0.33	56328704

Table 3.2. Tycho Brahe's apparent sizes of and average distances to the fixed stars, with resulting physical sizes *assuming a heliocentric cosmos* (E. R. = Earth Radius; S. D. = Solar Distance; E.V. = Earth Volume). Compare to table 3.1. As the Copernican heliocentric and Tychonic hybrid geocentric systems were mathematically identical regarding the Sun, Moon, and planets, values for those bodies are the same in both. Thus in a Copernican cosmos the physical sizes of the stars dwarf even the Sun (see fig. 3.9). As Tycho pointed out, a third magnitude star must be comparable in physical size to the orbit of Earth. Apparent diameter and distance values are from Brahe 1602, 424–31; also see Dreyer 1890, 190–91, and Thoren 1990, 302–7. Physical radius and volume are calculated from these, and so may differ slightly from Tycho's numbers, which suffer from rounding and typographical errors.

star would be the size of Earth's entire orbit. What's more, said Tycho, is that if the parallax turns out to be smaller than that one minute of arc, then the fixed stars would have to be larger still. Such immense stars at such immense distances were, Tycho said, absurdities of the Copernican hypothesis.[15]

Here was a weighty anti-Copernican argument. It was not based on Aristotelian ideas but simply on observation, measurement, and geometry. As Albert van Helden has put it, "Tycho's logic was impeccable; his measurements above reproach. A Copernican simply had to accept the results of this argument."[16] Only as long as one ignored the stars did the Copernican system offer a "more perfect heaven."[17] Riccioli would devote considerable time and effort to this powerful argument.

But this was not the only anti-Copernican argument of Tycho to have weight. He also argued that Earth's motion should be detectable through physical experiments—in short, we should see some evidence of Earth's

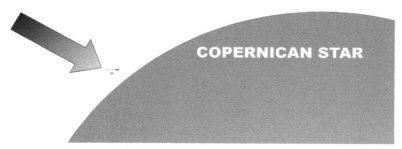

Figure 3.9.   The arrowed dots are fig. 3.7, reproduced to scale compared to Brahe's calculated relative size for a mid-size star in the Copernican universe (where the stars lie at vast distances, and thus must be enormous to explain their apparent sizes as seen from Earth). Brahe said the huge Copernican stars were absurd. From Graney 2013a.

motion in phenomena visible here on Earth. In discussing an exchange of letters that he had with the German Copernican Christoph Rothmann (ca. 1550–ca. 1600), Tycho brings up the subjects of a lead ball dropped from a tower and of identical cannons launching identical balls to the east and to the west. His view of the physics of these phenomena is a mixture of Aristotle and impetus: a cannonball has a natural motion downward owing to its weight, but the cannon can give it another, violent (not natural) motion resulting from the impetus provided by the explosive powder. But, says Tycho,

> that most violent motion indeed may impede the other motion (by which heavy bodies descend necessarily and naturally straight down), until after a long space is passed through and the violent motion will have diminished itself, and the ball by degrees settles into quiet and is able to reach Earth.[18]

This seems to echo Buridan, but whereas Buridan recognized the importance of air resistance on a moving body, Tycho did not believe that air, which he describes as "most fluid" and "so tenuous," could have a significant effect upon the motion of a heavy body.[19] The motion of a projectile— which in some cases does appear much like a rapid straight-line motion that dies out and transitions into a drop toward Earth (fig. 3.10)[20]—results

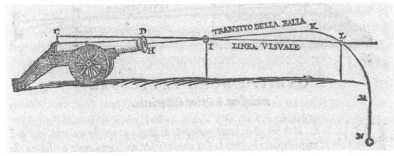

Figure 3.10. Top: The path of a projectile (from Tartaglia 1554, 16r), illustrating how a ball moves in a straight line under the impetus given to it by a cannon. The cannon impetus then dies out, allowing the ball to transition to its natural downward motion. Image credit: ETH-Bibliothek Zürich, Alte und Seltene Drucke.
Bottom: Trajectory of a low-density projectile moving through the air, calculated according to a modern understanding of projectile motion with air resistance (a similar graph appears in Graney 2013c, 413). Note the similarity between the two figures.

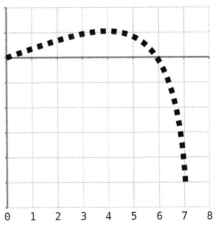

from, as Tycho sees it, the cannon giving the ball an initial, rapid motion (which diminishes of its own accord, an idea with which Buridan would not have agreed), and then from the ball's own natural downward motion, caused by its gravity (which does not diminish).

Following this way of thinking, a ball held atop a tower on a rotating Earth might indeed share in the tower's motion as the Earth turned, but when it was released, the impetus it received on account of the Earth's motion would begin to diminish, like the cannon ball's initial motion. The ball, as it fell, would therefore lag behind the tower, at least to some extent. Since points on Earth's surface would move to the east in the Copernican hypothesis—toward the rising Sun—the lead ball, being left behind, would deflect to the west as it fell. Likewise for cannon balls launched east and west: on a rotating Earth, both would tend to land west of the spots at which they would hit on an immobile Earth, with the result being that the east-launched ball would have a shorter range than the west-launched ball, if Earth rotated. As it is, said Brahe, objects actually fall straight down from high places—they do not deflect westward—and cannons actually do not have a greater range to the west than to the east. The best explanation for all this is that Earth does not move.[21]

The Copernican answer to such an argument was that everything on Earth's surface shares in Earth's rotational motion, with no diminishment, and therefore no effects are observable. Galileo would explain the Copernican "common motion" position nicely, by imagining the cabin of a ship:

Shut yourself up with some friend in the main cabin below decks on some large ship, and have with you there some flies, butterflies, and other small flying animals. Have a large bowl of water with some fish in it; hang up a bottle that empties drop by drop into a narrow-mouthed vessel beneath it. With the ship standing still, observe carefully how the little animals fly with equal speed to all sides of the cabin. The fish swim indifferently in all directions; the drops fall into the vessel beneath; and, in throwing something to your friend, you need throw it no more strongly in one direction than another, the distances being equal; jumping with your feet together, you pass equal spaces in every direction. When you have observed all these things carefully (though there is no doubt that when the ship is standing still everything must happen in this way), have the ship proceed with any speed you like, so long as the motion is uniform and not fluctuating this way and that. You will discover not the least change in all the effects named, nor could you tell from any of them whether the ship was moving or standing still. In jumping, you will pass on the floor the same spaces as before, nor will you make larger jumps toward the stern than toward the prow even though the ship is moving quite rapidly, despite the fact that during the time that you are in the air the floor under you will be going in a direction opposite to your jump. In throwing something to your companion, you will need no more force to get it to him whether he is in the direction of the bow or the stern, with yourself situated opposite. The droplets will fall as before into the vessel beneath without dropping toward the stern, although while the drops are in the air the ship runs many spans. The fish in their water will swim toward the front of their bowl with no more effort than toward the back, and will go with equal ease to bait placed anywhere around the edges of the bowl. Finally the butterflies and flies will continue their flights indifferently toward every side, nor will it ever happen that they are concentrated toward the stern, as if tired out from keeping up with the course of the ship, from which they will have been separated during long intervals by keeping themselves in the air. And if smoke is made by burning some incense, it will be seen going up in the form of a little cloud, remaining still and moving no more toward one side than the other. The cause of all these correspondences of effects is the fact that the ship's motion is common to all the things contained in it, and to the air also.[22]

Indeed, experiments like this on a ship will not reveal whether the ship is moving or not, for exactly the reasons Galileo describes. Tycho was wrong about motions diminishing of their own accord. And contrary to Tycho, and as Buridan said, air can have an effect on heavy objects. (Tycho apparently lacked experience with truly violent storms such as tornadoes.) The droplets fall from the bottle into the vessel beneath, whether the ship is at rest or in smooth motion, precisely because the horizontal motion that they have on account of the ship's motion does not diminish as they drop, and because the air is being carried along with the cabin, too. Likewise the balls dropped from the tower or launched from the cannon will not show the effects Tycho supposes. Galileo had a better grasp of the physics of motion than did Tycho.

After discussing the ship cabin example, Galileo goes on to say, "I am satisfied so far, and convinced of the worthlessness of all experiments brought forth to prove the negative rather than the affirmative side as to the rotation of the earth."[23] But Tycho had another cannon argument for the Copernicans, and this one skirted their common motion argument. Tycho's writing on this is long and convoluted. It is difficult enough in Latin, and even more so in English since English provides fewer signals than Latin regarding, for example, which pronoun goes with which word through the use of gender and case. But it contains a key point upon which Riccioli will later build:

And truly I add this, that if near the poles of the Earth, where the diurnal motion (if such might exist) comes to rest, the same experimentation by a gun [of cannons fired in different directions] might be made toward whatever part of the Horizon by the method previously said, the same might always come about, and if in the middle between each pole at the Equator, where the motion of the circumference of the Earth ought to be fastest: so that even in regard to any horizon at all, if a ball may be hurled by like method toward the east and west, it traverses the same space, which toward the South and North by like impulse hurled, when nevertheless the diurnal motion of Earth (if any might be there), it might respect west and east: truly South and North not likewise. Since then these may come about uniformly everywhere, indeed it is necessary that Earth rest uniformly everywhere.[24]

So what is Tycho struggling to say here? He is saying to imagine the experiment with cannons fired in different directions (which he had just finished discussing, and which we encountered a few paragraphs back) being conducted first at the poles of the Earth, and then at the equator. At the poles, he says, the Earth's diurnal rotation (*if* it exists, he says) produces no movement—a pole lies on the axis of a rotating Earth. Moreover, at the poles, regardless of which direction the cannon is aimed, the experiment must yield the same result. We might say that at the north pole, regardless of what point on the horizon a gun is aimed toward, its ball is always hurled southward. On the equator, by contrast, the motion caused by diurnal rotation is the fastest. The Earth measures about 25,000 miles in circumference, and the Earth rotates once in 24 hours, so at the equator the ground must be moving at 25,000/24 = 1042 mph, versus zero mph at the poles. Moreover, at the equator, direction matters. Tycho draws our attention to what it will mean if, at the equator, the cannon is fired east and west, and then north and south, and the flight of the balls is the same. At the equator the diurnal motion does not respect north and south in the same way as east and west. Tycho then says that since the flight of cannon balls actually happens in the same way everywhere, the Earth must be at rest.

While the sentence is convoluted and the argument is not precise, Tycho is hitting on a key point. The Copernican "common motion" argument does not really apply to the Earth. The Earth is not like a ship, in which everything moves together. Some points on Earth, such as the equator, move very fast; others, namely the poles, move not at all. Diurnal motion is consistent only when one moves east or west—along a parallel of latitude. Along a line of longitude, on the other hand, the diurnal motion changes, increasing in speed as one moves from a pole toward the equator, decreasing as one moves from the equator toward a pole. If one is in the northern hemisphere, a point on the ground a half a mile to one's north is moving more slowly than the point under foot, owing to Earth's diurnal rotation, and a point a half a mile to the south is moving faster. This is not a matter of Aristotelian physics or half-formed ideas about impetus. Like Tycho's star size objection, this is simply a matter of observation, measurement, and geometry. Tycho seems to have a gut feeling that somehow, if Earth rotates, the variation of that motion with latitude should produce detectable effects, and he concludes that since no such effects are observed, Earth must not rotate. As we shall see, his gut was onto something. Riccioli will develop Tycho's rotation objection from a vague notion into a solid anti-Copernican argument.

Tycho Brahe does not conform to the usual portrayal of anti-Copernican thinking (or nonthinking, as the case may be). A fine example of the usual portrayal can be found in the foreword that Albert Einstein wrote for Stillman Drake's translation of Galileo's 1632 *Dialogue Concerning the Two Chief World Systems, Ptolemaic and Copernican*, and which can be found in recent editions of that translation. Einstein characterizes anti-Copernican thinking as being a product of "anthropocentric and mythical thinking" rooted in a seventeenth-century "paralysis of mind brought on by the rigid authoritarian tradition of the Dark Ages," and as "opinions which had no basis but authority."[25] Tycho's anti-Copernicanism is often portrayed as being related to religion, or to the worldview of his time, or just to an inability on his part to take the next logical step forward. For example, the web page of the Tycho Brahe Museum (located at the site of Tycho's observatory on Hven) has a page on "Tycho Brahe's Worldview," which speaks of his geocentrism in terms of the established world order, mentions him saying the Bible had as much authority as science, and so forth.[26] It is not difficult to find statements, both in the popular press,[27] and in scholarly works,[28] about how he "could not bring himself" to accept the Copernican system. His powerful star size objection (one not at all based on authority or tradition or religion), an argument described centuries ago by the Dutch astronomer Christiaan Huygens (1629–1695) as Tycho's "principal argument" against the Copernican hypothesis, and once discussed even in encyclopedia articles, seems today to be something with which only specialists have acquaintance.[29] This is most unfortunate, for as we shall see, Riccioli was not the only anti-Copernican to root his arguments in the work of this great and most influential Danish astronomer.

4　Stars and Adventitious Rays

The stars, to the naked eye, present diameters varying from a quarter of a minute of space, or less, to as much as two minutes. The telescope was not then invented which shows that this is an optical delusion, and that they are points of immeasurably small diameter.

—from the article "Brahé, Tycho" in the 1836 *Penny Cyclopædia of the Society for the Diffusion of Useful Knowledge*

For the second chapter in a row we begin with the *Penny Cyclopædia*'s wonderful little discussion of Tycho's star size objection to the Copernican hypothesis. This time, however, we shall focus on the question of the telescope, for the *Penny Cyclopædia* summarizes what is commonly accepted to be the answer to that objection—namely, that the telescope showed that, in the words of the historian of science Stillman Drake, who has translated so much of Galileo's work into English,

> Fixed stars are so distant that their light reaches the earth as from dimensionless points. Hence their images are not enlarged by even the best telescopes, which serve only to gather more of their light and in that way increase their visibility.[1]

For this reason, Christine Schofield, in her discussion of the Tychonic hypothesis and its variants, tells us:

The absolute size of stars was no longer a problem, because the use of the telescope had made necessary a re-estimation of the true diameter of the fixed stars, which were now known to be far smaller than they appear to the naked eye.[2]

All this is owed to Galileo's assessment in his 1610 *Starry Messenger* that

the fixed stars are never seen to be bounded by a circular periphery, but have rather the aspect of blazes whose rays vibrate about them and scintillate a great deal. Viewed with a telescope they appear of a shape similar to that which they present to the naked eye, but sufficiently enlarged so that a star of the fifth or sixth magnitude seems to equal the Dog Star, largest of all the fixed stars.[3]

This notion that "the fixed stars appear as dimensionless points" has been repeated by the best of scholars, from Kepler, who cites Galileo,[4] to Van Helden, who cites Kepler,[5] to Grant, who cites Van Helden.[6]

Yet the fact of the matter is that fixed stars actually do *not* appear as dimensionless points. Galileo in fact did *not* view stars as being points. What Galileo relates in the *Starry Messenger* are the observations of a man who has been using a telescope for mere months. With a little more experience, he begins to refer to stars as spheres. As early as 1612–13, in his letters on sunspots, he wrote:

Stars, whether fixed or wandering, are seen always to keep the same shape, which is spherical.[7]

and

[A certain gentleman] thinks it probable that even the other stars are of various shapes and that they appear round only because of their light and their distance, as happens with a candle flame — and, he might well have added, with horned Venus. Such an assertion could not be proven false if it were not that the telescope shows us the shapes of all the stars, fixed as well as planets [wandering stars], to be quite round.[8]

Then in 1617, when he observed the double star Mizar with a telescope (a double star is one that appears to be a single star when seen with the unaided eye, but is found to be two stars very close together when seen with a telescope), he wrote in his observing notes that the telescope revealed Mizar to be two stars separated by 15 arc seconds (15/60 of an arc minute, or 15/3600 of a degree), one of which measured 3 seconds in radius, the other of which measured 2.[9] In his 1624 letter written in reply to Monsignor Francesco Ingoli (of whom we shall hear more in the next chapter), he wrote:

> I say that if you measure Jupiter's diameter exactly, it barely comes to 40 seconds, so that the sun's diameter becomes 50 times greater; but Jupiter's diameter is no less than ten times larger than that of an average fixed star (as a good telescope will show us), so that the sun's diameter is five hundred times that of an average fixed star[10]

and

> many years ago . . . I learned by sensory experience that no fixed star subtends even 5 seconds, many not even 4, and innumerable others not even 2.[11]

And finally, in his *Dialogue* of 1632 we see that

> the apparent diameter of the sun at its average distance is about one-half a degree, or 30 minutes; this is 1,800 seconds, or 108,000 third-order divisions. And since the apparent diameter of a fixed star of the first magnitude is no more than 5 seconds, or 300 thirds, and the diameter of one of the sixth magnitude measures 50 thirds [5/6 seconds] . . . , then the diameter of the sun contains the diameter of a fixed star of the sixth magnitude 2,160 times[12]

and

> [if a beam of wood mounted to serve as a reference for marking a star's position] is not large enough to hide the star, I shall find the

place from which the disc of the star is seen to be cut in half by the beam—an effect which can be discerned perfectly by means of a fine telescope.[13]

Thus it is clear that when Galileo observed stars with a telescope, he saw round disks, whose diameters he consistently measured to be a few seconds of arc. Galileo differed from Tycho only in terms of the apparent sizes he attributed to stars, not in whether he believed the stars to have an apparent size.

Galileo assumed that stars were approximately the same physical size as the Sun, and based on this assumption, calculated the distances to the stars, as seen in the letter to Ingoli,

> the sun's diameter is five hundred times that of an average fixed star; from this it immediately follows that the distance to the stellar region is five hundred times greater than that between us and the sun,[14]

and in the *Dialogue*,

> the diameter of [a star] of the sixth magnitude measures 50 thirds [5/6 seconds] . . . the diameter of the sun contains the diameter of a fixed star of the sixth magnitude 2,160 times. Therefore if one assumes that a fixed star of the sixth magnitude is really equal to the sun and not larger, this amounts to saying that if the sun moved away until its diameter looked to be 1/2160th of what it now appears to be, its distance would have to be 2,160 times what it is in fact now. This is the same as to say that the distance of a fixed star of the sixth magnitude is 2,160 radii of the earth's orbit,[15]

and in his Mizar notes, where he determines Mizar's component stars to be hundreds of times more distant than the sun.[16]

These sorts of distances simply did not comport to the lack of observable annual parallax, especially since Galileo also argued in the *Dialogue* that

> I do not believe that the stars are spread over a spherical surface at equal distances from one center; I suppose their distances from us vary

so much that some are two or three times as remote as others. Thus if some tiny star were found by the telescope quite close to some of the larger ones, and if that one were therefore very very remote, it might happen that some sensible alterations would take place among them.[17]

In this way, he said, even the fixed stars would appear in court to testify to Earth's annual motion.[18] Here Galileo is suggesting that in a double star one could observe, not the absolute annual parallax of the stars them-selves, but the difference in their parallaxes, which would cause the spac-ing between the two to change (see fig. 3.6 concerning stars stars σ and ω). Galileo's observations were astoundingly good,[19] and this method of de-tecting "differential parallax" would be very sensitive. Were Galileo's as-sumptions about the fixed stars being suns that were a few hundred or thousand times more remote than the Sun even remotely correct, this differential parallax method should have worked.[20]

But it did not.[21] Indeed, it is perplexing that, when Galileo made this claim in the *Dialogue* about what would happen "if" such a double star were to be found, roughly fifteen years had passed since he had observed just such double stars—Mizar, the Trapezium in Orion, and others—in around 1617. Another astronomer had suggested to him the differential parallax technique in 1611, so he certainly knew of the technique.[22] In all likelihood he was interested in double stars because of the possibility of detecting differential parallax.[23] But no differential parallax is observable in either Mizar or the Trapezium. Nor would it have been observable in any number of apparently suitable double stars that a thorough search for differential parallax would have turned up.[24]

What could explain the lack of parallax in the double stars? One ex-planation might be, of course, that the Earth did not move. Another might be that the Earth moved, but that Galileo's assumption that the stars were the size of the Sun was off by two orders of magnitude. In other words, the stars were two orders of magnitude farther away than Galileo said (which would explain the lack of parallax), and thus, by geometry, also two orders of magnitude larger than the Sun. This takes us back to Tycho's giant stars problem.

Galileo did not publish his double star observations, and they did not come to light until 2004.[25] He generally did not mention any problems that his telescopic observations of the stars could have produced for the

Copernican hypothesis. However, those observations make it clear that the telescope certainly did not solve Tycho's star size objection to that hypothesis. That objection was alive and well, despite the advent of the telescope. Galileo simply never addressed it.

But Galileo was not the only highly skilled observer with a good telescope. Simon Marius (1570–1624) of Ansbach, Germany, claimed that the appearance of stars through a telescope supported the Tychonic, not the Copernican, hypothesis. Literature about Marius and his work is scarce. The literature that exists typically notes that he observed the Jovian system; that he claimed that he independently discovered the moons of Jupiter, thereby incurring Galileo's ire (the names of Jupiter's four Galilean moons—Io, Europa, Ganymede, and Callisto—come from Marius);[26] and that he observed the object we now know as the Andromeda galaxy with his telescope, recording numerous details about it.[27]

But Marius was a skilled observer. In discussing Andromeda in his 1614 book, *The World of Jupiter*, Marius describes

> a fixed star or kind of star of remarkable form that I came upon and saw by means of a telescope the night of 15 December of the Year 1612. In the whole heaven I am not able to discover another such star. But it is near the third and northernmost star in the belt of Andromeda. Without the instrument it is discerned as a kind of little quasi-cloud in that spot; with the instrument no distinct stars are seen (like in the nebula of Cancer, and other nebulous stars), but whitish rays, which where they are nearer the center there grow brighter. The light is dull and pale in the center. It occupies almost the quarter part of a degree in diameter. The luster appears almost like if a candle shining through translucent horn were to be discerned from far off. It appears not unlike to that Comet in the Year 1586.[28]

This description is testimony to Marius's ability as an observer and to the quality of his telescope. Only those familiar with the view of this object through a telescope (who know that, even with modern telescopes far superior to what was available to Marius, many observers cannot describe the Andromeda galaxy so clearly) will fully appreciate what Marius was able to accomplish with such a small instrument. Further testimony comes

from Anton Pannekoek, who has pointed out that Marius, from his observations of Jupiter's moons, derived better values than did Galileo for their
periods of revolution and other orbital elements.[29] And the well-known
historian of astronomy J. L. E. Dreyer has noted that Marius observed the
disks of stars telescopically.[30]

Marius comments about observations of stellar disks, and how they
supported the Tychonic hypothesis, in *The World of Jupiter*. The comments are not detailed, and they are not well known. They are not included in the commonly available English translation of his book.[31] Marius writes that

> I obtained an instrument, through which not only the planets, but
> also all the more conspicuous fixed stars I observed, are seen round
> (especially the great dog, the small dog, and the brighter stars in
> Orion, Leo, Ursa Major, etc.). Before that time I had never happened
> to see this. I am truly surprised Galileo did not see this with his most
> excellent instrument. Indeed he writes in his *Starry Messenger*, the fixed
> stars appear in no way restricted by a circular periphery—something
> which certain persons since have considered grounds of the greatest of
> arguments. In truth, by this statement itself they confirm the Coperni-
> can world system: it is on account of the immense Copernican dis-
> tance of the fixed stars from Earth that their globe shape cannot be
> perceived from Earth by any method at all. Since truly now it may be
> most certainly established, that by this telescope on the Earth even
> the fixed stars are seen to be circular, this line of argument surely
> falls, and the contrary is plainly built up: specifically, that the sphere
> of the fixed stars is by no means removed from the Earth by such an
> incredible distance as the speculation of Copernicus produces. Rather,
> such is the segregation of the fixed stars from the Earth, by the harmo-
> nious Tychonic[32] ordering of the spheres of the heavens, as the struc-
> ture of those bodies may nevertheless be distinctly seen to be the
> shape of a circle by this instrument.[33]

Marius notes additional evidence for the Tychonic system in the
moons of Jupiter (whose observed motions he could reconcile with his
calculations only if they circled Jupiter while Jupiter circled the sun).[34] He

adds that these ideas will require further discussion and explanation (which he does not provide). Finally, Marius concedes to Galileo that the stars shine by their own light—they are distinct in appearance from the planets, being notably more intense in brilliance.[35]

Marius was not the only astronomer to write about how the appearance of stars when seen through a telescope undermined Copernicus. Peter Crüger (1580–1639) of Gdańsk wrote in his 1631 *Astrophysical Delicacies*,

> How should they [the stars of varying magnitudes] through a telescope all appear confusedly as mere points? For have not *Galileo* and others, by means of the telescope, discerned stars of a further six magnitudes otherwise invisible to natural sight? So surely, one must admit that, even through a telescope, the stars have an apparent diameter. Take only a tiny star, one whose diameter is merely a quarter of a minute [15 seconds], or 1/120th that of the Sun ... From *Kepler's* calculation that there are *60,000,000 Earth-radii* between the center and the firmament, it therefore follows ... that the true diameter of such a tiny star is *2,181 Earth-diameters* .... Now, also according to *Kepler*, the Sun has a diameter of 15 Earth-diameters. So from this it follows that the diameter of the tiny star I mentioned must be more than 145 times greater than the diameter of the Sun, and likewise (by the authority of *Euclid*, book 12, proposition 18) the body of the same star must be 3,048,624 times as great as the Sun. Indeed, moreover, if a tiny star had a visible breadth even of a mere 60th of a minute [one second], then according to the above hypothesis its true thickness would be more than nine times that of the Sun, and its volume 730 times greater. I therefore do not understand how the Pythagorean or Copernican *Systema Mundi* can survive.[36]

Crüger's writing is particularly interesting because he notes that even a star of the smallest size—one whose diameter is but a single second of arc—poses a challenge to the Copernican system. Both Marius and Crüger comment on people claiming stars are mere points. Kepler is one person who made that claim,[37] and Crüger makes use of Kepler's work.

Marius and Galileo both observed that stars seen through the telescope show distinct disks (it is unclear whether Crüger observed the disks himself). Marius, Galileo, and Crüger all note that the more prominent stars show a larger disk. Galileo, for example, says in the *Dialogue* that first-magnitude stars measure 5 arc seconds in diameter, while sixth magnitude stars measure a sixth that size. Marius says in *The World of Jupiter* that the disks are most noticeable in the brighter stars. Crüger speaks of a variety of disk sizes.

Marius declares these disks to support a Tychonic world system. Crüger says that the Copernican system cannot survive. Galileo does not speak to the problems his stellar disk observations cause for Copernicus, but should have.[38]

Then what of the telescope showing that the apparent sizes of stars as seen by the naked eye are "an optical delusion"? What of the telescope showing that stars are points of immeasurably small diameter? What of the business about how, thanks to the telescope, the absolute sizes of stars were no longer a problem, and Tycho's star size objection was dissolved?

The answers to these questions are not simple. The appearance of stars, whether "fixed" or "wandering" (that is Mercury, Venus, Mars, Jupiter, and Saturn—the planets), can be perplexing. Consider two early seventeenth-century telescopes that are otherwise identical, except that one magnifies the church tower down the street by 20 times, while the other magnifies it by 40 times. Compared even to the most modest of modern telescopes, such as a quality small telescope that might be given as a gift to a child or used by a backyard bird watcher, our two seventeenth-century instruments are very small. Their apertures (the hole that allows light into the lens) each measure an inch or less in diameter. Their fields of view are small. The images they produce are lacking in detail.

Imagine these two telescopes being turned toward the Moon. Recall that the Moon measures, as seen by the naked eye, approximately half a degree, or 30 arc minutes, or 1800 arc seconds, in diameter. To the eye, 180 Moons could be lined up edge to edge in the 90 degrees between the zenith and the horizon. These telescopes will enlarge the Moon by 20 and 40 times, respectively. However, if the 20 power and 40 power telescopes are used to measure the Moon's apparent diameter, each will show it to be approximately half of a degree (although each will probably measure it to respectively greater precision). This is because, while each telescope enlarges the Moon, it also enlarges everything else, including all the space between the zenith and the horizon. Each telescope will still show that 180 Moons would fit edge-to-edge between the zenith and the horizon. Thus, in the case of the Moon, the telescope confirms the measurements of the naked eye, and of Brahe's nontelescopic instruments.

Yet this is not the case with a wandering or fixed star. For example, seen nontelescopically, with the naked eye, Venus and Jupiter each appear as very prominent stars. As we saw in chapter 3 (table 3.1), with nontelescopic instruments Tycho measured the apparent diameter of Venus to be 3.25 arc minutes (3 minutes and 15 seconds of arc) when Venus is 1 S. D. (solar distance) from Earth; he measured the apparent diameter of Jupiter to be 2.75 arc minutes (2 minutes and 45 seconds) when 5.2 S. D. from Earth. Yet, whereas in the case of the Moon the telescope confirms the measurements of the naked eye and of Brahe's instruments, in the case of Venus and Jupiter it does not. The 20 power telescope will *not* enlarge Venus and Jupiter by 20 times compared to Tycho's measurements. It will *not* confirm his 3.25 and 2.75 minute apparent diameters of Venus and Jupiter. Neither will the 40 power telescope.

Instead, our two telescopes show these two bodies to appear much smaller than, and different from, what the naked eye sees. Through each telescope, Venus and Jupiter both appear as round bodies with apparent diameters of less than a minute. Venus is obviously round, but does not show a full disk; viewed over the course of many months it waxes and wanes through a set of phases similar to the moon, and it changes in size significantly as its relative distance from Earth changes (fig. 4.1). Jupiter shows a full round disk, but with dark belts on it. It has moons. It changes in size somewhat over the course of many months as its distance from Earth changes (the effect is less than what is seen in Venus, owing to Jupiter being farther away). However, regardless of these complexities, when used to observe these planets the two telescopes will produce results consistent with each other. The planetary apparent diameters measured by the two telescopes, and the planetary appearances seen through the two telescopes, may disagree with the diameters and appearances determined nontelescopically, but they agree with each other. Venus will appear twice as large in the 40 power telescope as in the 20 power telescope, and so will Jupiter. The 40 power telescope will probably return more precise measurements of their diameters, but the two telescopes will be in agreement nonetheless.

Figure 4.1.   The appearance of Venus changing over time, as seen by Galileo using a telescope (Galilei 1623, 217). Image courtesy History of Science Collections, University of Oklahoma Libraries.

If the 20 power telescope reveals a crescent Venus, so will the 40 power. If the 20 power shows three moons to the east of Jupiter and one to the west, the 40 power will agree. It is with the naked eye, and with nontelescopic instruments, that the telescopes disagree.

If our two telescopes are next turned to the fixed stars, those stars will appear as round bodies like Jupiter, but considerably smaller. Tycho, with nontelescopic instruments, measured first magnitude stars as being 2 minutes, or 120 seconds, in diameter; Galileo (as we have seen), Riccioli, the Dutch astronomer Martinus Hortensius (also known as Martin van den Hove [1605–1639]),[39] and the Polish Johannes Hevelius[40] (1611–1687) all measured such stars as having diameters of less than 20 seconds. Through both the 20 and the 40 power telescopes the stellar disks are so small that precise measurement is difficult, but the stars clearly appear as Galileo and Marius have stated: as round bodies, with the brighter stars being larger than the fainter ones. A prominent star such as Arcturus has a larger diameter than a barely visible star such as Alcor. The two telescopes will produce consistent results when used to observe fixed stars, just as they did with Jupiter and Venus. The stellar diameters measured by the two telescopes, and the stellar appearances seen through the two telescopes, will disagree with the diameters and appearances recorded nontelescopically, but they will generally agree with each other. Fixed stars will appear twice as large in the 40 power telescope as in the 20 power telescope, and the 40 power telescope will probably return more precise measurements, but the two will be in agreement nonetheless. As was the case with the planets (wandering stars), it is with the naked eye, and with nontelescopic instruments, that the telescopes disagree.

Galileo, in his *Starry Messenger*, proposed a theory to explain this disagreement. He said that in the case of a bright body with small apparent size, a telescope reveals the true size of the body by stripping away what he called "adventitious rays" caused by moisture in the eyes. The idea was that a small, intense light source produced a sort of glare that hid the body of the source:

> When stars [fixed or wandering] are viewed by means of unaided natural vision, they present themselves to us not as of their simple (and, so to speak, their physical) size, but as irradiated by a certain fulgor and as fringed with sparkling rays, especially when the night is far advanced. From this they appear larger than they would if stripped of those adventitious hairs of light, for the angle at the eye is determined not by the primary body of the star [fixed or wandering] but by the brightness which extends so widely about it.[41]

A telescope, he said, "removes from the stars [fixed and wandering] their adventitious and accidental rays, and then it enlarges their simple globes."[42] In his 1632 *Dialogue* Galileo discusses this theory further, by way of relating the appearance of a small but brilliant object to that of a candle seen from a distance at night, and again uses the theory to explain why the apparent sizes of stars and planets are so much larger when seen with the naked eye, and when measured with nontelescopic instruments, than when observed with a telescope.[43] He discusses at length why various astronomers measured the fixed and wandering stars to be too large (italics added):

> And truly I am quite surprised at the number of astronomers, and famous ones too, who have been quite mistaken in their determinations of the sizes of the *fixed as well as the moving* stars, only the two great

luminaries [the Sun and Moon] being excepted. Among these men are al-Fergani, al-Battani, Thabit ben Korah, and more recently Tycho, Clavius, and all the predecessors of our Academician [Galileo]. For they did not take care of the adventitious irradiation which deceptively makes the stars [fixed and wandering] look a hundred or more times as large as they are when seen without haloes. Nor can these men be excused for their carelessness; it was within their power to see the bare stars [fixed and wandering] at their pleasure, for it suffices to look at them when they first appear in the evening, or just before they vanish at dawn. And Venus, if nothing else, should have warned them of their mistake, being frequently seen in the daytime so small that it takes sharp eyesight to see it, though in the following night it appears like a great torch. I will not believe that they thought that the true disk of a torch was as it appears in profound darkness, rather than as it is when perceived in lighted surroundings; for our lights seen from afar at night look large, but from near at hand their true flames are seen to be small and circumscribed.[44]

The adventitious rays theory seemed to work. Riccioli concurred with it. The telescope, he said, in "exposing the disks of the stars and scraping off the adventitious ringlets of the rays" is more trustworthy than nontelescopic instruments such as Tycho's.[45] Riccioli provided thorough reports on the details visible on the Moon and planets only through the telescope—including all those lunar features he named in the lunar map in the *New Almagest* (recall from chapter 1)—as well as Jupiter's cloud bands (and their occasional changes) and even the earliest report of Jupiter's Great Red Spot.[46] His reports on Jupiter, for example, included both his own observations and the observations of other observers (fig. 4.2). All skilled observers with good telescopes saw the Jovian cloud bands.

And so it seems obvious what a telescope does: it removes adventitious rays, enlarges the disk of the fixed star or wandering star (planet) being observed, and provides a trustworthy image of that body's true appearance. To the eye Venus appears to be a significant fraction of the Moon's size, and featureless in appearance, but that appearance is false because of the adventitious rays. Venus's true size, and true appearance, are those revealed by the telescope. The same holds true for Jupiter, Arcturus,

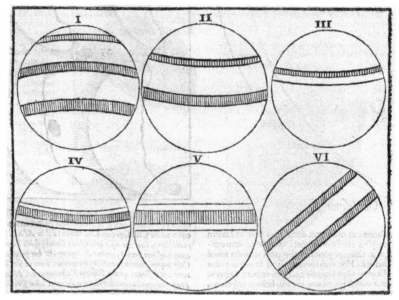

Figure 4.2.   Representations of Jupiter's cloud bands as seen by various observers, from the *New Almagest* (Riccioli 1651, 1:486). Image credit: ETH-Bibliothek Zürich, Alte und Seltene Drucke.

and Alcor. Thanks to the ray-removing telescope, we know, as Galileo said, that all stars, be they fixed or wandering, are seen always to keep the same shape, which is spherical.[47] We know that even if someone asserts that it is probable that different stars are of various shapes and that they appear round only because of their light and their distance, as happens with a candle flame (and as happens with Venus as regards its phases), that assertion can be proven false, because the telescope shows us the shapes of all the stars, fixed as well as planets, to be quite round.[48]

But there is a problem with this assessment of what a telescope does: it is wrong. Four hundred years of learning about and using telescopes has taught us that, yes, Venus's true size and true appearance are those revealed by the telescope. And, yes, the same holds true for Jupiter. But the same does *not* hold true for Arcturus and Alcor. The telescope indeed shows the shapes of the planets to be quite round. But, appearances to the contrary, it reveals nothing about the fixed stars. Unbeknownst to Galileo,

Marius, Riccioli, and other telescopic observers in the first half of the seventeenth century, there was a strange difference between wandering stars (planets) and fixed stars. In the case of the planets, the "simple globes" the telescope revealed and enlarged indeed *were* the true physical bodies of the planets. But in the case of the fixed stars, the "simple globes" the telescope revealed and enlarged *were not* their true physical bodies. Those simple globes that Galileo and Marius saw when they looked at the fixed stars through a telescope were in fact false and spurious.

Why? The answer is not at all simple. It is perhaps best explained by borrowing an idea from Galileo. Imagine seeing a candle burning in the dark, quite some distance away—perhaps two hundred yards or so. At such a distance, what one sees is only the star-like appearance of a gleam of light. The apparent size of the gleam of light is entirely spurious, having nothing to do with the actual size of the flame. The apparent size depends mostly on how much light is reaching the eye, so the gleam will grow smaller and larger as the candle burns less or more brightly.

Why is the apparent size spurious? Because the physics of light—principally a phenomenon known as "diffraction," by which waves of light passing through any opening are distorted by that opening—imposes some inherent limits on the ability of the eye to see detail. This ability is also known as *resolving power*. No one can read this page of type from one hundred feet away, because the apparent size that the print would have at that distance is too small. It falls below the resolution limit of the eye. No one's eye can "resolve" the print from that distance. For the candle flame burning a few hundred yards away in the dark, the eye can see the light, but it cannot resolve the flame.

Now imagine walking toward the flame. So long as the apparent size the flame would have is below the resolution limit of the eye, you will continue to just see the gleam of light. However, that gleam may grow larger because the light from the flame grows more intense as you draw nearer to it. The spurious apparent size is increasing! At some point, however, you will draw close enough to the flame that its apparent size is large enough for the eye to resolve, and the true body of the flame itself becomes visible.

The Moon and Sun are large enough that the eye can resolve them—the eye can reveal their true bodies. But stars, whether they be fixed stars or planets (wandering stars), are too small for the eye to resolve, and

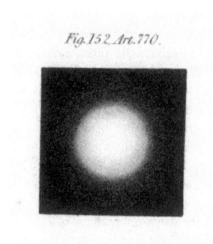

Fig. 15 2. Art. 770.

Figure 4.3. A star as seen through a small aperture telescope (see Herschel 1828, 491, and Plate 9). This appearance of a sphere of measurable size is entirely spurious—an artifact of diffraction. However, early telescopic astronomers took such telescopic images to be the physical bodies of stars (Grayson and Graney 2011). This disk would be smaller than the smallest sketch of Venus in fig. 4.1. Image credit: ETH-Bibliothek Zürich, Alte und Seltene Drucke.

instead a clear-eyed observer who looks at the stars sees small, entirely spurious, disks of light, which appear to be of smaller or larger "bigness" or "magnitude" (recall from chapter 2), depending on how bright the star is. When Tycho Brahe measured the apparent sizes of celestial bodies with his nontelescopic instruments, he was measuring the true bodies of the Sun and Moon, and measuring the spurious disks of everything else.

But just as the eye has a resolution limit, so does a telescope. The Moon and Sun are large enough that the telescope can resolve them and reveal their true bodies—and so are Mercury, Venus, Mars, Jupiter, and Saturn. But a fixed star is still too small for the telescope to resolve. Moreover, a telescope such as Galileo or Marius used—a good-quality but very small aperture telescope—produces a spurious disk that is a remarkably convincing stand-in for the true body of a fixed star (fig. 4.3), and that can be magnified and measured just like the true bodies of Jupiter or Mars.

Not only were the spurious star disks convincing, but they were certainly *not* the sort of problem that telescope users would be looking for. As we shall see in chapter 10, it would be very difficult to determine that the globe that a telescope revealed when trained on a fixed star like Arcturus was spurious, whereas the globe a telescope revealed when trained on a wandering star like Jupiter was true, and that the size difference between Arcturus and Alcor that a telescope revealed was spurious, whereas the size difference between Jupiter and Mars that a telescope revealed was

true. And when the challenge was to convince people that what the telescope revealed was real (even today, people who get their first look at Saturn through a good-quality modern telescope will often suggest that the
image is faked, as the contrast between the dot of light they see in the sky
and the beautiful world they see in the telescope is just too great), who
would even think to look for such an insidious and bizarre problem?

Thus while it is true that *eventually* the telescope showed the sizes of
stars to be spurious—an "optical delusion"—and *eventually* it showed
the apparent sizes of stars to be points of immeasurably small diameter, at
first the telescope did nothing of the sort. Moreover, the absolute sizes of
stars remained a problem, and Tycho's star size objection was not dissolved. Rather, the telescope was thought to reveal the true disks of fixed
stars in the same way it revealed the true disks of wandering stars (that is,
of planets), and the star size objection remained. This is an important but
forgotten reality. We must remember this reality as we proceed forward in
our investigation of Riccioli and the scientific case against the Copernican
world system.

5   Science against Copernicus, God's Starry Armies for Copernicus

All have said the stated proposition to be foolish and absurd in Philosophy; and formally heretical, since it expressly contradicts the sense of sacred scripture in many places, according to the quality of the words, and according to the common exposition, and understanding, of the Holy Fathers and the learned Theologians.[1]

—statement regarding the Copernican hypothesis, from a committee of eleven consultants for the Roman Inquisition, 24 February 1616

In the period between the advent of the telescope and Riccioli's *New Almagest*, Tycho Brahe's objections to the Copernican system retained currency in the world system debate. The star size objection can be found in writings from this period by Johann Georg Locher, Francesco Ingoli, and Philips Lansbergen. Locher and Ingoli were both anti-Copernicans whose ideas Galileo criticized in some detail in his writings. Both referenced Tycho when arguing for Earth's immobility. Lansbergen, on the other hand, was a prominent Copernican who adapted Tycho's star size objection into a religious argument for the superiority of the Copernican system. Locher, Ingoli, and Lansbergen illustrate not only that Tycho's objections remained part of the debate, but they also illustrate a surprising aspect of that debate: anti-Copernican reliance on "scientific" arguments to support their views, and Copernican reliance on "religious" arguments to support theirs.

Johann Georg Locher (1592–1633) was a student of Christopher Scheiner, who in turn was a Jesuit and professor of mathematics at Ingolstadt. In 1614 Locher published a book entitled *Mathematical Disquisitions on Controversial and Novel Astronomical Topics*. Galileo in his *Dialogue* discusses Locher's book extensively. William Donahue describes Locher and Scheiner as influential Jesuit scientists who sought to renovate astronomical ideas rather than build them anew, and who accepted telescopic discoveries and applied their own explanations to them.[2] Indeed, in the *Disquisitions*, Locher discusses and elaborately illustrates various telescopic discoveries, such as the rugged lunar surface and spots on the Sun (fig. 5.1). He includes a fold-out illustration of how the telescopically discovered phases of Venus indicate that it circles the Sun (fig. 5.2). But he does not accept the Copernican system.

Locher provides six key arguments against Copernicus. The first of these makes a partial reference to matters of religion, namely, that the Copernican system requires that many passages of sacred scripture, and many astronomers' common ways of speaking, be twisted in a preposterous sense.[3] He elaborates on this colorfully:

> Because of the arrangement of the parts of the Universe and the imagined motion of Earth in this hypothesis, the Sun, Mercury, and Venus are below, and Earth is above. Heavy bodies absolutely and naturally ascend—light bodies descend. Christ the Lord ascended to hell, and descended to the heavens, at which time he approached the Sun. When Joshua commanded the Sun to stand still, actually the Earth stood still, or the Sun moved contrarily to earth. When the Sun dwells in Cancer, the Earth runs through Capricorn, and generally the winter zodiac signs bring about the summer, and the summer the winter. The stars do not rise and set to Earth, but Earth to the stars—the stars' rising becomes the Earth's setting, the stars' setting becomes the Earth's rising. In short, the whole course of the World, as it were, is inverted.[4]

Yet, acknowledges Locher, "The Copernicans may answer arguments like these adequately but convolutedly."[5] In other words, arguments about how the Copernicans invert the meaning of astronomical or scriptural texts carry little weight. Copernicans *can* answer them, even if the answers seem convoluted. However, says Locher, "they will not be able to satisfy so

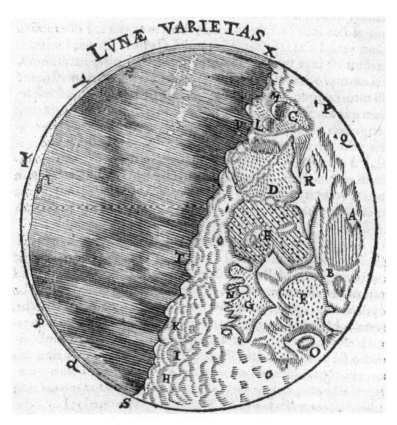

Figure 5.1.   Illustrations of the Moon (top) and Sun (bottom) from Locher's 1614 *Disquisitions*, pages 58 and 65. Note the craters on the Moon and the spots on the Sun. Images credit: ETH-Bibliothek Zürich, Alte und Seltene Drucke.

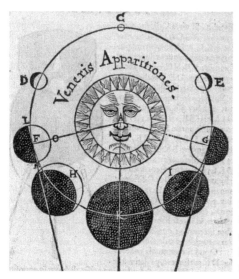

Figure 5.2. Diagram from Locher's 1614 *Disquisitions* (page 76) showing how the telescopically observed phases of Venus demonstrate that it circles the sun. Image credit: ETH-Bibliothek Zürich, Alte und Seltene Drucke.

well the arguments that follow."[6] He has other arguments he says the Copernican *cannot* answer.

And what is the next key argument Locher provides? The star size argument. Locher says that the Copernican motion of the Earth makes any first-magnitude star well larger than the whole orbit of the Earth, and makes any small star just as large.[7] Locher cites Tycho, providing books and page numbers, in his discussion of this argument.[8]

The remaining four key arguments consist of one argument about the inelegance of the Copernican system (the vast distance of the stars serves no purpose, other than to explain the lack of parallax) and three about the Copernican system being in conflict with the Aristotelian physics of motion.[9] Thus to Locher, the question of scripture makes up but half of a lightweight argument against Copernicus. The weighty anti-Copernican arguments are questions of science: star sizes, star distances, and physics. Yet, in light of the telescopic discoveries of Galileo, says Locher, a system is needed that can accommodate Venus circling the Sun—a hybrid geocentric system like that of Tycho Brahe (see fig. 5.3).[10]

Just two years after Locher's *Disquisitions*, in January of 1616, Monsignor Francesco Ingoli (1578–1649) addressed Galileo with an essay entitled "Disputation Concerning the Location and Rest of Earth against the System of Copernicus," which echoed Locher to some extent. In this essay

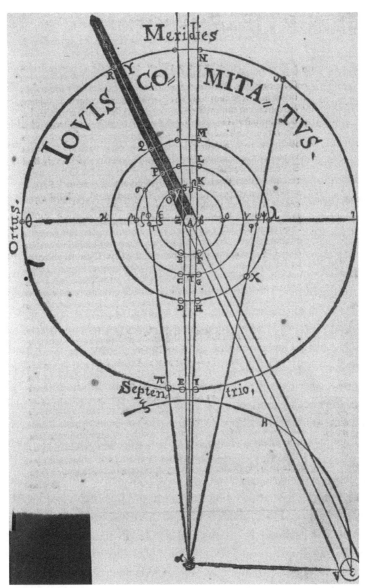

Figure 5.3. The Jovian system as illustrated in Locher's 1614 *Disquisitions* (page 81). Locher shows the Jovian moons, the Jovian shadow, the points where the moons will either pass in front of or behind the Jovian disk as seen from Earth with a telescope, and the points where the moons are hidden by the Jovian shadow. However, note that Jupiter and its company are illuminated by a Sun (lower right) that itself is circling the Earth (at lower center). This and the previous two figures show how Locher accepted and clearly illustrated telescopic discoveries, yet did not interpret them as supporting the Copernican system. Within *Disquisitions* he says why not—the problem of star sizes under the Copernican hypothesis being one prominent reason. Image credit: ETH-Bibliothek Zürich, Alte und Seltene Drucke.

Ingoli presents a greater number and wider variety of arguments against the Copernican hypothesis than Locher does in *Disquisitions*. Ingoli describes the arguments as mathematical, physical, and theological. But, like Locher, he puts emphasis on the questions of science: mathematical (i.e., astronomical) and physical arguments. In his closing paragraph, Ingoli suggests that Galileo focus on those arguments, rather than on the theological arguments, and then only on the "more weighty" among them. Like Locher, Ingoli also relies on the work of Tycho Brahe. Many of the essay's mathematical and physical arguments reference Tycho's book *Astronomical Letters*, often including page numbers.

Galileo believed Ingoli's essay to have been influential in the infamous condemnation of the Copernican hypothesis by a committee of consultants for the Roman Inquisition in February 1616. Thus Ingoli's emphasis on weighty mathematical and physical arguments (that is, "scientific" arguments) rather than on theological arguments, his reliance on the work of Tycho, and even his inclusion of theological arguments, which he then deemphasizes, all suggest an overlooked aspect of the Inquisition's opposition to Copernicus. Unsurprisingly, among those arguments of Tycho that Ingoli cites are the cannon at the equator and poles argument, and of course, the star size objection.

Ingoli was a well-connected clergyman who had engaged Galileo in debates at the home of Lorenzo Magalotti (1584–1637). After one of their oral debates, the two agreed to put their arguments into writing, with Ingoli doing so first and Galileo responding. Thus Ingoli produced the essay.[11] Maurice Finocchiaro writes that this essay was "one of the most complete, intellectually ambitious, and historically significant" of the anti-Copernican critiques produced at the time of Galileo's telescopic discoveries. It was not published, but it was circulated widely. Finocchiaro continues that Ingoli "had probably been commissioned by the Inquisition to write an expert opinion on the controversy" and that this opinion "provided the chief direct basis for the recommendation by its committee of consultants that Copernicanism was philosophically untenable and theologically heretical."[12] Galileo was of the opinion that the essay played an important role in in the rejection of the Copernican hypothesis by church authorities in Rome. In a reply to the essay that he wrote to Ingoli in 1624, he notes that Ingoli's arguments "were not lightly regarded by persons of authority who may have spurred the rejection of the Copernican opinion."[13] He worries about Ingoli's essay having "circulated in various foreign nations" and about people thinking "that the rejection of Copernicus's opinion was based on a belief that you [Ingoli] were right."[14] While the essay may have circulated, it was not published until the end of the nineteenth century, in Antonio Favaro's *Opere di Galileo*. Translations into English are scarce. A translation is provided here, in appendix A (part 1), along with notes on technical aspects of the essay's arguments and Galileo's 1624 responses to them (part 3).

Ingoli's essay may indeed have been important in the rejection of the Copernican hypothesis by church authorities, but at the time he wrote the essay, that rejection was yet to occur. The Dominican friar Niccolò Lorini (1544–ca. 1617) had filed a complaint against Galileo with the Inquisition in Rome in February 1615. That complaint was investigated by the Inquisition during 1615, and Galileo had gone to Rome in December 1615 in hopes of clearing his name and preventing the condemnation of the Copernican hypothesis.[15] It was in the month following Ingoli's essay, on 24 February 1616, that the Inquisition's committee of consultants issued their opinion condemning the Copernican hypothesis as "foolish and absurd in philosophy." This was promptly followed by a meeting on 26 February between Galileo and Cardinal Robert Bellarmine (1542–1621) in which Bellarmine purportedly warned Galileo that "the doctrine attributed to Copernicus . . . is contrary to Holy Scripture and therefore cannot be defended or held." Then on 5 March the Congregation of the Index (the department within the Vatican in charge of book censorship) issued a decree rejecting the Copernican hypothesis as "false" and "altogether contrary to Holy Scripture."[16] Thus Ingoli's essay just preceded the rejection.

Ingoli opens the essay with a "preface" in which he makes reference to the discussions before Magalotti regarding the Copernican hypothesis, noting that Galileo has been the defender of Copernicus, while he (Ingoli) has been given the role of defending the other side—bringing forth arguments to support the hypothesis of the "old mathematicians" and to tear down the Copernican assumption. He says he has agreed to his role willingly, because he is grateful and honored to be debating with and learning from educated men such as Galileo.

The body of Ingoli's essay consists of anti-Copernican arguments, which we will discuss presently. However, Ingoli closes the essay with a short paragraph that is a particularly interesting feature of the essay and that illuminates the arguments. Thus it needs to be presented in its entirety here first:

> These [arguments] complete this disputation. Let it be your choice to respond to this either entirely or in part—clearly at least to the mathematical and physical arguments, and not to all even of these, but to the more weighty ones. For I have written this not toward attacking your erudition and doctrine (most notable to me and to all men both inside the Roman Curia and outside), but for the investigation of the truth, which you profess yourself always to search for by all strength, and in fact so suits a mathematical talent.[17]

Indeed, when Galileo wrote his response to Ingoli in 1624, he only responded to Ingoli's mathematical and physical arguments. He did, however, respond to every one of those, in great detail. Galileo's reply, despite the absence of responses to the theological arguments, is over twenty thousand words in length; Ingoli's essay has a length of approximately three thousand words.[18]

A technical discussion of each mathematical and physical argument is found in appendix A, part 3. Ingoli presents Galileo with five physical arguments against Copernicus, most of which deal with the fact that the Copernican hypothesis conflicts with the Aristotelian physics of motion. He presents Galileo with thirteen mathematical anti-Copernican arguments. These cover a wide range of subjects, and are of uneven quality, but it is here that we find Tycho's arguments about the cannon at the poles and equator, and about star sizes.

The star size argument features prominently in Ingoli's essay, yet he does not put it forth as one of the thirteen mathematical arguments per se. Rather, he puts it into a special discussion that follows the second and third mathematical arguments, both of which have to do with matters related to parallax. These two arguments cite two different noticeable effects that would be visible in the fixed stars were the Earth not in the center of the universe (as it would not be, were it orbiting the Sun). The Copernican answer to such arguments, as we have seen, is that the orbit of Earth is of negligible size compared to the distance to the stars.

At this point Ingoli brings in the star size question, saying,

> Nor does the solution [to the two parallax arguments] entirely satisfy
> by which is said: the diameter of the circle of the orbit of Earth in
> comparison to the vast distance of the eighth orb from us to be made
> so small [as to yield an effect too small to measure].[19]

He then cites a page of Tycho's *Astronomical Letters* as well as his calculations showing how distant the stars would have to be in the Copernican
system for annual parallax to be too small to detect. Here is the very first
time that Ingoli mentions Tycho Brahe in the essay—when he brings out
Tycho's toughest anti-Copernican argument. Ingoli continues:

> Such a truly great distance not only shows the universe to be asymmet-
> rical, but also clearly proves ... the fixed stars to be of such size, as they
> may surpass or equal the size of the orbit circle of the Earth itself.[20]

For these reasons, Ingoli says, the parallax arguments cannot be dismissed
by the Copernican assertion that the stars are so distant. And neither, we
may presume, can the other arguments that involve the effects of Earth's
motion seen in the stars (the seventh and eighth mathematical arguments
are such arguments).

In his 1624 reply, Galileo answered that the telescope showed stars to
measure much smaller in diameter than Tycho had measured with nontelescopic instruments: seen with a telescope, the stars were revealed to be
less than a tenth the size Tycho had measured.[21] But as we saw in chapter 4, in 1614 Simon Marius had stated in his book *The World of Jupiter*
that telescopic observations of the stars argued for a Tychonic world system, and by 1624 Galileo's own (unpublished) telescopic observations of
stars had apparently shown Tycho's star size objection to be valid still.

Remarkably, Ingoli actually suggests to Galileo the solution to Tycho's
argument. Ingoli suggests, in essence, that the apparent sizes of the stars are
spurious. In the "such a truly great distance" quote above, Ingoli says that
the great distance either proves the fixed stars to be of enormous size, or, he
says (at the point of the ellipses between "clearly proves" and "the fixed
stars"), it proves that the fixed stars, on account of their great distance, do

not function like other celestial bodies seen from Earth. He offers as an analogy that the Sun has less effect the further it is from the zenith (in winter versus in summer). In his 1624 reply, Galileo interprets this statement as Ingoli saying that if the stars were vastly distant it would destroy their ability to affect things on Earth (perhaps in an astrological sense),[22] but Ingoli is clearly speaking in regards to the apparent sizes of stars:

> Such a truly great distance not only shows the universe to be asymmetrical, but also clearly proves, either the fixed stars to be unable to operate in these lower regions, on account of the excessive distance of them; or the fixed stars to be of such size, as they may surpass or equal the size of the orbit circle of the Earth itself.[23]

A loose paraphrase might read, "either the stars must be huge, or something doesn't work like we think it does." As we saw in chapter 4, Ingoli was right—as regards stars, something indeed did not work like they thought it did.

Ingoli devoted less space to Tycho's cannon at the equator and poles argument. Citing the appropriate pages from the *Astronomical Letters*, Ingoli notes Tycho's argument

> concerning the bombards discharged from the east into the west and from the north into the south, particularly concerning those discharged near the poles, where the movement of Earth is slowest. For, given the diurnal motion of Earth, the most apparent differences would be observed, whereas nevertheless no differences are observed.[24]

This is Ingoli's sixth mathematical argument. Almost all the mathematical arguments cite Tycho in some way.

Ingoli also includes four theological arguments in his essay, although, as seen earlier, he suggests Galileo bypass them in favor of the mathematical and physical ones, which Galileo does in his 1624 reply. These arguments concern how the words in scripture describe a geocentric structure to the heavens, how hell (in the center of the Earth) should be at the farthest point from heaven, how Joshua commanded the sun to stand still, and how geocentric language is used in certain prayers. Interestingly, Riccioli would include within his 1651 *New Almagest* a discussion of 126 pro- and anti-

Copernican arguments much lengthier than Ingoli's (discussed in chapter 7 in this volume). Only two of the 126 would be of a religious or scriptural or theological nature. One of those two would be the same argument about the placement of heaven and hell that Ingoli presents to Galileo. Riccioli would dismiss this argument because, he would say, hell is a place defined in comparison to God's heaven and the world of men; whether the Earth has motion or not is irrelevant.[25] The theological arguments also feature Cardinal Robert Bellarmine prominently. Ingoli specifically mentions him in conjunction with the second and fourth theological arguments.

But theological arguments in the essay about hell or scriptural language are outnumbered more than four to one by mathematical and physical arguments—by "scientific" arguments (in Locher's *Disquisitions* they are outnumbered five and one-half to one-half). These scientific arguments, and in particular the more weighty ones, are those Ingoli suggests Galileo focus on.

They are not all equally weighty. At least one is easily dismissed (see appendix A, part 3, argument M1). But the others are better, some are much better, and those taken from Tycho Brahe's book *Astronomical Letters* would carry a significant weight owing simply to the authority of their illustrious author.

Ingoli cites Tycho repeatedly in the essay—far more frequently than he cites older sources such as Aristotle or Ptolemy. Despite Ingoli's remark in the preface about his defending the "old mathematicians," it is not the ancient works of Aristotle and Ptolemy that Ingoli brings to bear against Copernicus, but the recent works of Tycho, the leading astronomer of the time. It is Tycho, not Aristotle and Ptolemy, that Ingoli sees as being the competition for Copernicus.

And as we saw in chapter 3, Tycho's arguments not only carried the weight of the authority of the leading astronomer of the time, but they were also—in light of the knowledge of the time—strong, valid arguments. Tycho's challenging cannon-at-the-equator-and-poles argument, for example, was one that Galileo failed to adequately answer in his 1624 reply to Ingoli, and presumably Galileo could not have answered it in 1616 either. And, of course, Ingoli included in his essay a special discussion of Tycho's strongest argument: star sizes. Ingoli recognized what was the weightiest argument in the essay, probably because Tycho had recognized it as such. As we shall see, Riccioli, in his *New Almagest* analysis of 126 pro- and anti-Copernican arguments, dismissed the vast majority of the arguments on either side as being inconclusive (and sometimes even fatuous)—with the exception of two classes of arguments. First and foremost of these two were arguments involving the apparent sizes of stars. And then, secondarily, were arguments related to the matter of the cannon at the equator and poles. These arguments could not be answered (at the time), and the inability to answer them would be explicitly recognized by Riccioli.

In light of these strong scientific arguments, the purpose of the theological arguments (downplayed by Ingoli and bypassed by Galileo) seems twofold. One purpose seems to be to convey a sense of Cardinal Robert Bellarmine's views on the matter, as two of the theological arguments mention him. The other seems to be to emphasize how the words of scripture fit a geocentric universe. Ingoli notes in discussing the third theological argument that, "in explaining Sacred Writings the rule is to always save the literal sense, when it can be done," echoing Augustine and Aquinas. Bellarmine had emphasized this point some months earlier in April 1615, stating in a letter that if solid evidence for the Copernican system were found, then the literal sense of scripture would have to give way to a different interpretation:

> I say that if there were a true demonstration that the sun is at the center of the world and the earth in the third heaven, and that the sun does not circle the earth but the earth circles the sun, then one would have to proceed with great care in explaining the Scriptures that appear contrary, and say rather that we do not understand them than that what is demonstrated is false. But I will not believe that there is such a demonstration, until it is shown me. Nor is it the same to demonstrate that by supposing the sun to be at the center and the earth in heaven one can save the appearances, and to demonstrate that in truth the sun is at the center and the earth in heaven; for I believe the first demonstration may be available, but I have very great doubts about the second, and in the case of doubt one must not abandon the Holy Scripture as interpreted by the Holy Fathers.[26]

Bellarmine had apparently held this sort of view about scripture and astronomy for decades.[27] It seems to have been common in Jesuit circles, but the French Jesuit Honoré Fabri (1607–1688) was apparently the first to publish it, writing in 1661 that

> nothing hinders that the Church may understand those Scriptural passages that speak of this matter in a literal sense, and declare that they should be so understood as long as the contrary is not evinced by any demonstration[28]

and continuing on to say that if some demonstration of the validity of the Copernican hypothesis were found, the Church would not scruple to declare that those passages are to be understood in a figurative sense.[29]

Thus Ingoli seems to be stating that the theological arguments only come into play if the science allows—that is, "when it can be done." If the science is clearly against scripture, then the interpretation that "Scripture may speak following our manner of understanding" must be invoked. It is understood that scripture does not assert that the value of pi is three when it describes a round pool of water whose circumference is three times its diameter.[30] Nor does scripture make the scientific assertions that the world is flat with four corners or that no seeds are smaller than mustard seeds.[31] Ingoli can suggest that Galileo pass over the theological arguments and focus on the more weighty of the scientific arguments, because the crux of the matter is scientific.

But science in 1616 is not clearly against scripture: according to Tycho Brahe, the top expert in the field, the science does not support the Copernican system. Indeed, as we saw in chapter 3, Tycho believed it to be absurd. Thus the literal sense of scripture can still be retained, and the Copernican system, in contradicting that literal sense, is "heretical."

Ingoli's reliance on Tycho Brahe, and Ingoli's suggestion that Galileo ignore his theological arguments, as well as Locher's use of Tycho's ideas and his minimal focus on scriptural arguments, again raise a question regarding the manner in which opposition to the Copernican system is often depicted. Namely, such opposition is typically portrayed as a matter of adherence to Aristotle or to religion. It is not often depicted as a matter of valid scientific arguments, supported by detailed references to works by a leading scientist. In discussing Tycho in chapter 3, we noted Albert Einstein's references to "anthropocentric and mythical thinking" and so forth.

Yet Ingoli's essay, written a matter of weeks before the condemnation of the Copernican system by the Inquisition's consultants and likely the basis of that condemnation, rests neither on references to religion nor on references to Aristotle. Rather, Monsignor Ingoli's anti-Copernican essay rests largely on references to the mathematical and physical anti-Copernican arguments of Tycho Brahe, the preeminent astronomer of the time. The same can generally be said of Locher's work.

If anti-Copernicans like Locher and Ingoli relied on Tycho Brahe's arguments to support hybrid geocentrism, how did Copernicans respond to those arguments in defending heliocentrism? As noted in chapter 3, Albert van Helden has stated that, insofar as Tycho's star size argument was concerned, both the logic it used and the measurements upon which it was based were unquestionably solid, so Copernicans just had to accept the argument. And so they did. Some Copernicans simply claimed the giant stars to be evidence of the power of God. After all, an infinite and omnipotent God can make stars as large as desired. Thus stars as gigantic as Copernican stars, and even a cosmos as immense in size as the Copernican cosmos, shrink to nothing by comparison to God. Riccioli complained in the *New Almagest* about Copernicans using this solution to the star size problem. A fascinating, dramatic, and colorful illustration of this reliance on a religious argument to solve a scientific problem with the Copernican system is found in the work of the Dutch Copernican Philips Lansbergen (1561–1632). Lansbergen stands in marked contrast to Locher's and Ingoli's reliance on scientific arguments to oppose Copernicus.

Lansbergen was not the first Copernican to invoke God's power in regard to the star size question. Indeed, Copernicus himself invoked God when discussing the stars in his 1543 *On the Revolutions*. Noting that Earth's motion is not reflected in the fixed stars, he writes,

> This proves their immense height, which makes even the sphere of the annual motion, or its reflection, vanish from before our eyes. For, every visible object has some measure of distance beyond which it is no longer seen, as is demonstrated in optics. From Saturn, the highest of the planets, to the sphere of the fixed stars there is an additional gap of the largest size. This is shown by the twinkling lights of the stars. By this token in particular they are distinguished from the planets, for

there had to be a very great difference between what moves and what does not move. So vast, without any question, is the divine handiwork of the most excellent Almighty.[32]

In 1576 the English Copernican astronomer Thomas Digges (1546–1595) also used language that spoke of the stars in terms of God's handiwork when he published his English translation of portions of *On the Revolutions* together with a sketch (fig. 5.4) of the Copernican universe under the heading "A perfit description of the Cœlestiall Orbes." Digges could advocate this sort of view with true passion. In his 1573 *Mathematical Wings or Ladders*, he argues that the brilliant new star or "nova" that had recently appeared in the heavens presented a chance to prove Copernicus correct through parallax measurements. Digges describes the nova as being a terrifying and wonderful miracle of God—the Star of Bethlehem returned again. And indeed, its brilliance (it could be seen by daylight) and its auspicious location (on the vernal equinox and passing directly overhead as seen from England) must have been awe-inspiring. In the concluding paragraph of *Mathematical Wings*, Digges issues a rousing call to arms for astronomers. This Monster of the Heavens, he says, may provide the long-awaited chance to emend and correct the old geocentric system with that of "divinely inspired Copernicus—him of more than human ingenuity."[33] Truly, says Digges, I see no better method by which we might understand the amazing work of God; this is why men are given eyes! Those blessed with astronomical and mathematical skills must join in the Olympic battle and make the measurements and do the calculations. We have, says Digges, the Mathematical ladders needed to scale the towers of the Heavens, to measure the distances and structures of the Universe, and to investigate this portentous Star that once announced the birth of Christ to the Magi. Thus we may indubitably testify about a stupendous Miracle of God to others, and indeed to all to whom is not given to lift up their faces from the Earth; so all might know the mighty works of God, to whom alone is due all Praise, Honor, and Glory, for all time.[34]

In his "Perfit Description," which he published shortly after *Mathematical Wings*, Digges again discusses the heavens in religious terms. Within his sketch of the infinite universe he writes this description of the universe of stars:

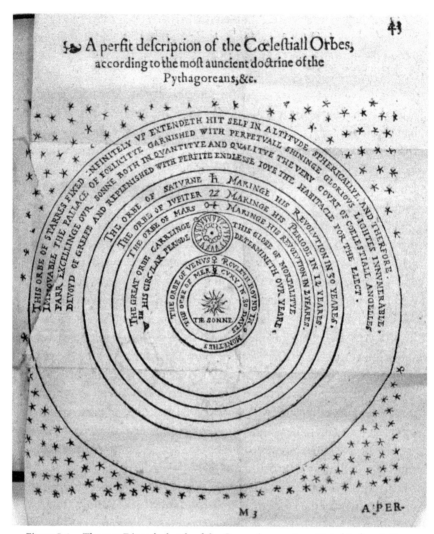

Figure 5.4. Thomas Digges's sketch of the Copernican system. Note his description of the starry heavens as the "palace of felicity" garnished with innumerable glorious lights "far excelling our sun both in quantity and quality," the court of the angels and the dwelling of the elect. Such language is reflected in the writings of various Copernicans, including Christoph Rothmann and Philips Lansbergen. Image courtesy History of Science Collections, University of Oklahoma Libraries.

THE PALLACE OF FOELICITYE GARNISHED WITH PERPETUALL SHININGE
GLORIOUS LIGHTES INNUMERABLE. FARR EXCELLINGE OUR SONNE BOTH
IN QUANTITYE AND QUALITYE THE VERY COURT OF COELESTIALL AN-
GELLES DEVOYD OF GREEFE AND REPLENISHED WITH PERFITE ENDLESSE
IOYE THE HABITACLE FOR THE ELECT.[35]

In his translation of Copernicus, Digges adds the following commentary:

Herein can we never sufficiently admire this wonderful and incom-
prehensible huge frame of God's work proponed to our senses.... We
may easily consider what little portion of God's frame our Elementary corruptible world is, but never sufficiently of that fixed Orb garnished with lights innumerable and reaching up in Spherical altitude
without end.... And this may well be thought of us to be the glorious
court of the great God, whose unsearchable works invisible, we partly
by these his visible, conjecture; to whose infinite power and majesty,
such an infinite place, surmounting all other both in quantity and
quality, only is convenient.[36]

Another Copernican to accept the idea of titanic stars and to invoke
this sort of imagery in explaining them was Christoph Rothmann, with
whom Tycho corresponded (recall from chapter 3). When Tycho put the
star size argument to Rothmann, Rothmann responded with:

[W]hat is so absurd about [an average star] having size equal to the
whole [orbit of Earth]? What of this is contrary to divine will, or is
impossible by divine Nature, or is inadmissible by infinite Nature?
These things must be entirely demonstrated by you, if you will wish
to infer from here anything of the absurd. These things that vulgar
sorts see as absurd at first glance are not easily charged with absurdity, for in fact divine Wisdom and Majesty is far greater than they understand. Grant the Vastness of the Universe and the Sizes of the stars
to be as great as you like—these will still bear no proportion to the
infinite Creator. It reckons that the greater the King, so much more
greater and larger the palace befitting his Majesty. So how great a
palace do you reckon is fitting to GOD?[37]

Brahe viewed Rothmann's response itself as absurd, arguing,

> On what is such an assertion based? Where in nature do we see the
> Will of God acting in an irregular or disorderly manner? In nature
> where all things are well ordered in all ways of time, measure, and
> weight? In nature where there is nothing empty, nothing irrational,
> nothing disproportionate or inharmonious. Consider [the vast dis-
> tance] between Saturn and the fixed stars in the Copernican hypothe-
> sis. Consider those same fixed stars each being as large as the whole
> Orbit of Earth (and some larger still) and thus dwarfing the Sun, the
> luminary and center of motion for all the planets. These are the same
> fixed stars that are noted as the least of the heavenly lights in the ac-
> count of the Creation of the World. This is empty, irrational, dispro-
> portionate, and inharmonious. Is such a disproportionate universe
> reasonable?[38]

Brahe goes on to remark that by Divine providence such a nongeometri-
cal, asymmetrical, disorderly, and most unworthy method of philosophiz-
ing will go away.[39] Indeed, Rothmann would eventually give up his Copernicanism and adopt Brahe's version of geocentrism.[40]

But, Lansbergen stands above Digges and Rothmann as a particularly
striking and colorful example of a Copernican invoking the power of God
in response to the star size argument. Lansbergen's discussion of this matter is found in his 1629 *Bedenckinghen, op den daghelijcksen, ende iaerlijck-
sen loop vanden aerdt-cloot*, one of the first public defenses of the Copernican system. *Bedenckinghen* was translated into Latin by Martin van den
Hove and published as *Commentationes in Motum Terrae Diurnum, et Annuum*, or *Considerations on the Diurnal and Annual Motion of the Earth*.
Lansbergen had argued that a geocentric cosmos implied absurdities—
the speeds of heavenly bodies whirling daily about an immobile Earth
were not to be believed.[41] Giant stars, on the other hand, were believable.
In *Considerations*, Lansbergen envisions three heavens—in part owing
to 2 Corinthians 12:2: "I know a man in Christ above fourteen years ago
(whether in the body, I know not, or out of the body, I know not), such
a one caught up to the third heaven." The first heaven is the planetary
heaven, extending up to the orbit of Saturn. The second heaven is the
starry heaven, extending from Saturn up to the orb of the fixed stars. The

third heaven is the empyreal heaven, the throne of God and the place reserved for the elect. The first and second heavens we can see; the third we cannot.[42] Lansbergen gives his discussion on the second, starry heaven the title "Concerning a True Description of the Second Heaven."[43]

He begins this true description by attempting to illustrate how much larger the second heaven is than the first. Just as the Earth is nearly a point in comparison to the first heaven (as defined by the sphere of Saturn), he says, so the orbit of the Earth is nearly a point in comparison to the second heaven (as defined by the sphere of stars). "Whence it follows, individual Fixed Stars, to be equal in size to the Orbit of Earth—a thing that almost overcomes both human comprehension and belief."[44] But such a great size is consistent with celestial motions, he says. Saturn requires 30 years to complete its course, and the orbit of Saturn measures 9.9 times the diameter of the orbit of Earth. Therefore, since the stars complete their course in 26,000 years (the precession of the equinoxes), the starry orb should be proportionately larger. It will dwarf the size of 14,000 Earth Radii that Tycho assigned as the radius of the whole universe.

But, says Lansbergen,

this Size of the Sphere of the *Fixed Stars* may be better appropriate to the Character of God, than the *Tychonic* size, because according to this great size, *God* is more correctly perceived to be immense, even infinite. For who may argue the *Second Heaven* is like to infinite unless God is in fact infinite? Since *He may fill up the Heaven and Earth,* Jeremiah 23:24, and *the Heaven of Heavens may not hold Him,* 1 Kings 8:27. So vast an *Atrium,* which he may have built on to his *Palace,* is indeed more glorious to the supreme Majesty of the Divine, than if he might have fashioned less.[45]

Anyone who thinks such a vast heaven is absurd, says Lansbergen, should keep in mind three things. First, God may have made this heaven most vast, but its size is proportionate to the size of the luminaries (the stars) that he has placed in the highest location of the firmament. Were the second heaven to be smaller, the light of the stars would overwhelm the light of the bodies of the first heaven. "Therefore He has had the reckoning, of *Number,* of *Measure,* and of *Weight,* in the construction of each Heaven: and He has united them mutually by absolute perfection."[46]

The second thing to keep in mind is that while the second heaven is exceedingly vast, it is nothing compared to God. God is infinite and immense. Nothing on Earth or in the heavens has any size compared to God. Lansbergen cites Isaiah 40:15–17, which notes that the nations are nothing—a drop in the bucket, dust in the balance—to God. And lest the reader think that perhaps Earthly comparisons leave open the possibility that the vast heavens might indeed compare to God, Lansbergen invokes Isaiah 40:26,

> Lift up your eyes on high, and see who hath created these things: who bringeth out their host by number, and calleth them all by their names: by the greatness of his might, and strength, and power, not one of them was missing.

Lansbergen is quick to add that we should not be overawed by the stupendous size of the heavens:

> It is necessary we may perceive these rightly; lest clinging to the stupendous size of the Heavens, and of the Celestial Bodies; we may hold them in place of the Creator, like the *Heathens:* but we may honor and we may adore only *God*, as the *Creator requiring to be praised into all ages, amen.* Romans 1:25.[47]

The third thing to keep in mind is that,

> although the space between the Orb of *Saturn*, and the *Fixed Stars* may be enormous, it is not empty, as *Tycho Brahe* with followers has thought. Rather it is full with the Creations of *God.* . . . occupied everywhere by the greatest multitude of invisible Creations.[48]

From here Lansbergen proceeds into an extensive discussion of angels and demons, which he supports by numerous references to scripture. Thus, he says, "it follows that the *Second Heaven* is not a vacuum, but is completely full of these Spiritual Creatures,"[49] in particular the angels coming and going from the third heaven, carrying out the business of God. Lansbergen mentions Jacob's Ladder from Genesis 28, with angels ascending and descending it: "Therefore as the Ladder seen by *Jacob*, was nowhere empty, but everywhere full of *Good* Angels; Thus the *Second Heaven* is never empty, but so full of the descending or ascending Armies of *Jehova*, as no place may be left void."[50] He continues on this theme for several paragraphs, citing scripture to support his ideas: "Concerning a True Description" extends for five pages, from page 27 through the end of page 31, and this discussion of how the second heaven is full of angels occupies more than a full page.

Upon wrapping up the angels discussion, Lansbergen turns to the outer limit of the second heaven—the eighth sphere, or orb of the fixed stars. This is the vault of heaven. If one considers an egg by way of comparison, he says, the first (planetary) heaven is like the yolk, the second heaven is like the white, and the sphere of the fixed stars is like the shell. The fixed stars are huge beyond measure, he says, although on account of their extraordinary distance from us they appear truly small. Their God-given purpose is not so much to illuminate Earth as to indicate the seasons of the year, and to be useful to farmers and sailors. But their purpose also is to illuminate the second heaven, just as God has placed the Sun in the center of the first heaven to illuminate it. But fixed stars are so much brighter than the Sun individually, and are so much more numerous, that their brightness exceeds that of the Sun by as much as the size of the second heaven exceeds that of the first.

> Indeed, if the Light of one *Sun* may be so excellent that the eyes may not consider it without injury; how much stronger will be the light of the so many and so much brighter Bodies that are gathered together in the *Second Heaven*? Therefore rightly the Apostle has said, 1 Timothy 6:16, *God to dwell in inaccessible light.* For if the splendor of the *Eighth* Sphere (which is the Atrium of the *Divine Palace*) may be so illustrious; how strong and inaccessible will be the Radiance of the *Habitacle of the Divine* Majesty itself?[51]

Lansbergen drives this point home with still more scriptural references.

Lastly, Lansbergen considers the nature of the fixed stars themselves. To this Copernican, the stars are nothing less than the giant warriors of God:

The *Fixed Stars* God has placed at a distance in the *Firmament*, as an *Army*. . . . Indeed truly they are visible *Armies of God*; by which the *God of Hosts [Deus Zebaoth]*, that is, of *Armies*, strongly does battle day to day.[52]

Again Lansbergen cites Isaiah 40:26. He also cites the canticle of Deborah and Barack, which mentions the stars doing battle (Judges 5:20), and he cites Joshua ordering the heavens to stand still:

For as *God* is *God of Armies*, which He commands; thus the *Stars* are the battling *Armies of God*, which obey and execute His Orders.[53]

We will properly understand that the stars are the warriors of God, says Lansbergen, if we consider four things. First, he says (turning the star size problem into support for his opinion), is the size of fixed stars:

For every single one exceeds not only the *Earth*, but also the *Sun*, and generally the whole *Orb of the Earth*. Moreover without doubt *God* has added Powers to them, proportional to Size. They may stand in Heaven by rank as the form of Giants, and of mighty Warriors, without any motion. For no one is able to move *Earth*, except *God* alone who has built it (Haggai 2:21), so no one knows how to stir up these immense and very powerful Bodies; except *God* the Creator of them. Therefore they are truly great and powerful *Armies of God*; through which, *He brings about anything He will wish* (Psalm 115:3).[54]

The second thing to consider is the vast multitude of the fixed stars, says Lansbergen. An army is considered to be stronger or weaker based on how many soldiers it has, he writes, and he cites the armies of the Israelites and of Xerxes and of Tamerlane, all of which numbered over a half million men. And while Ptolemy and Brahe may have only been able to observe some thousands of stars, he remarks, the famous astronomer Galileo Galilei revealed with the telescope that the stars are vastly more numerous, and indeed uncountable: "Therefore it is again made clear that the *Army of the Fixed Stars* is not only great; but also greater than all armies which have ever been on Earth: so they may be called the *Armies of the Lord* rightly by merit."[55]

The third thing to consider, according to Lansbergen, is the order found in the fixed stars. God did not make the stars like the sands of the sea, so that one star is just like another. Rather, God assigned them different sizes and brightnesses, as noted by both the Apostle Paul (in 1 Corinthians 15:41: "For star differeth from star in glory") and by Ptolemy—six classes of size, or "magnitude." God also amassed the stars into diverse constellations, forms, and figures. Thus they may be easily recognized, just like the units of an army on the battlefield are discerned by their specific positions and through particular banners and standards.[56]

Finally, Lansbergen says, the motion of the fixed stars is fully consistent with the artful advance of an army, where the skillful commander arranges the army by ranks and files, and preserves that order even in battle. Just this sort of motion is presented in the motion of the fixed stars from day to day. Indeed, God has prepared the line of battle of the fixed stars, and they have followed that line from the founding of the world, never departing a hair's breadth from their positions (recall from chapter 2 the unchanging nature of the heavens).

Lansbergen says the stars may also be considered the mighty Palace Guard of God. These guards endure we mortals stumbling onto the grounds of the palace, but they will gladly open the way when Our Lord Jesus Christ appears, so that Christ may raise us gloriously to himself, and stand us before the Throne of God—in the Third Heaven (which Lansbergen then goes on to discuss).[57]

Thus in Philips Lansbergen's *Considerations* we see a particularly interesting and colorful example of how a Copernican dealt with Tycho Brahe's star size objection to the Copernican system—an example that contrasts with the work of the anti-Copernicans Locher and Ingoli. Lansbergen, following Copernicans like Thomas Digges and Christoph Rothmann, incorporated Brahe's giant stars into the Copernican system by appealing to the power of God. Lansbergen declared the giant Copernican stars, which Brahe had viewed as an absurdity, to be an illustration of God's power—the army of God—a view Lansbergen thought was supported by scripture. The work of Locher, Ingoli, and Lansbergen set the stage for Riccioli and his *New Almagest*. Riccioli followed in the footsteps of Locher and Ingoli, writing an anti-Copernican work that relied on scientific arguments, especially the star size argument, and that criticized Copernicans for appealing to God's power to solve that argument.

6　Jesuits on the Tower

We have arrived at the point where we can discuss in some depth Giovanni Battista Riccioli and his efforts to weigh geocentrism against heliocentrism. After all, we are now familiar with the night sky. We are familiar with Tycho Brahe's star size objection to heliocentrism (his "principal argument"[1] against Copernicus) and with the trajectory of that objection over time. We are familiar with the telescope and what it revealed about stars. And we are familiar with the responses Copernicans offered to the star size objection. Let us now get to know Riccioli better—through his science.

Specifically, let us examine his falling body experiments. These are often cited as the first precise experiments to determine the acceleration due to gravity.[2] Why look at this work first? Because these experiments are relatively straightforward: Riccioli drops balls from high places and measures the times of their falls. His discussion of these experiments is equally straightforward. This straightforwardness leaves room for Riccioli's personality to show through. Through his own description of the falling body experiments, we see his creativity, his obsession with detail and accuracy, his ability to draw others into his work and form a "research team," and his concern with providing all the information needed for others to reproduce and test his work. We also see Riccioli's interest in sharing interesting results, even when those results happen to disprove his ideas and to prove the ideas of Galileo (whom, as we will see, he frequently criticizes).

Edward Grant has remarked that, while Riccioli was a geocentrist, unlike some geocentrists who "were not scientists properly speaking but natural philosophers in the medieval sense using problems in Aristotle's *De caelo* and *Physics* as the vehicle for their discussions, Riccioli was a technical astronomer and scientist."[3] Riccioli's work certainly reflects a scientist's interest in observing the real world—a scientist who was very, very interested in getting the observations right. In his description of Riccioli's efforts to calibrate a pendulum, Alexandre Koyré provides an entertaining illustration of Riccioli's concern for detail and accuracy. Riccioli would use pendulums to measure time in the falling body experiments, and Koyré is describing, in dramatic style, the calibration of a pendulum that would measure out precise seconds, and that could also serve as a standard against which quicker pendulums could be calibrated (fig. 6.1):

> For six consecutive hours, from nine o'clock in the morning to three o'clock in the afternoon, he counts (he is aided by the R. P. Francesco Maria Grimaldi) the oscillations. The result is disastrous: 21,706 oscillations instead of 21,660. Moreover, Riccioli recognizes that for his aim the sundial itself lacks the wanted precision. Another pendulum is prepared and "with the aid of nine Jesuit fathers," he starts counting anew; this time—the second of April 1642—for twenty-four consecutive hours, from noon to noon: the result is 87,998 oscillations whereas the solar day contains only 86,640 seconds.
>
> Riccioli makes then a third pendulum, lengthening the suspension chain to 3 feet, 4.2 inches. And, in order to increase the precision even more, he decides to take as a unit of time not the solar, but the sidereal day. The count goes on from the passage through the meridian line of the tail of the Lion (the twelfth of May 1642) till its next passage on the thirteenth. Once more a failure: 86,999 oscillations instead of 86,400 that there should have been.[4]

Here Koyré is describing the use of stars to mark precise time—in particular the use of an instrument to mark the moment when a star is on the "meridian" (the halfway point between rising and setting). The Sun is on the meridian at noon. The time between consecutive crossings of the meridian by a star is a "sidereal day," or 86,400 "sidereal seconds." The time be-

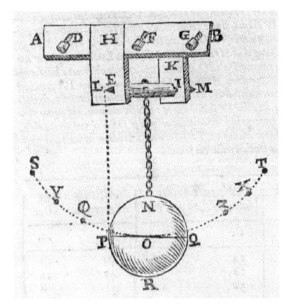

Figure 6.1.
Pendulum diagram
from the *New Almagest*
(Riccioli 1651, 1:84).
Image credit: ETH-
Bibliothek Zürich, Alte
und Seltene Drucke.

tween consecutive solar noons is four minutes longer (recall from chapter 2 the four-minute difference between the Sun and the stars), at 86,640 sidereal seconds. Koyré continues:

> Disappointed yet still unbeaten, Riccioli decides to make a fourth trial, with a fourth pendulum, somewhat shorter this time, of 3 feet, and 2.67 inches only. But he cannot impose upon his nine companions the dreary and wearisome task of counting the swings. Father Zeno and Father F. M. Grimaldi alone remain faithful to him to the end. Three times, three nights, the nineteenth and the twenty-eighth of May and the second of June 1645, they count the vibrations from the passage through the meridian line of the Spica (of Virgo) to that of Arcturus. The numbers are twice 3,212 and the third time 3,214 for 3,192 seconds.[5]

There is far less drama in Riccioli's words on this matter than in those of Koyré.[6] Riccioli provides information on the first pendulum, which at 21,706 vs. 21,600 is in error by a mere 0.2 percent, but adds that, "because the approach of the shadow toward the hour lines cannot be exactly discerned without error of some Seconds," he also includes the information on the other pendulums, which were measured over a period of years using different timing methods.[7] These others are each less accurate than the 21,706/21,600 pendulum, but all are accurate to well under 2 percent. Riccioli wants to leave no room for doubt concerning the quality of his timing pendulums.

With this in mind, let us turn to Riccioli's report on falling bodies.[8] For the most part, I will let Riccioli speak for himself here. Under the title "The Group of Experiments about Unequal motion of Heavy Bodies descending faster and faster in the Air, by which they come nearer to the end to which they tend,"[9] he starts by describing a few experiments that qualitatively indicate that a ball dropped from a greater height strikes the ground with more force, while supplementing these descriptions with examples from literature that might be familiar to his readers:

> Let a ball of wood or bone or metal fall from a height of 10 feet into an underlying bowl, and attend to the ringing arising from the percussion. Then let that same ball fall from a height of 20 feet, and indeed you will perceive a far greater and more extensive sound poured out. Next lift up the bowl to a height of 10 feet and into that let drop the same ball from an altitude of 10 feet above the bowl. Indeed you will perceive a ringing like at first. Therefore that ball in the second fall has acquired a greater impetus because of the drop from a greater height, than from the smaller heights in the first and third falls. And in the second fall the ball has gained more impetus in the second half of the journey down than in the first half, by as much as its downward velocity will have increased in the latter 10 feet of the fall, than in the former 10 feet.[10]

Riccioli also notes how water poured from a greater height produces greater noise, and that Cicero reports that people near the cataracts of the Nile have been deafened because of the noise produced by the falling water. He then proceeds to discuss the impact generated by falling objects:

> Place your hand below a ball while someone lets it fall from an altitude of 10 feet. Indeed you will experience the lightest impact. But if the same ball is let fall from an altitude of 50 feet or greater, your

hand will perceive some pain from the impact: therefore a greater impetus is derived from a higher fall. Poor Aeschylus[11] felt this greater impetus that I have described, from the turtle that the Eagle dropped onto his head; and the stupid bird herself felt no doubt concerning what would happen. Elpenor[12] sensed it in falling from the tower. So Ovid writes in the third book of Tristia

> Who falls on level ground—though this scarce happens—
> so falls that he can rise from the ground he has touched,
> but poor Elpenor who fell from the high roof
> met his King a crippled shade.[13]

And from this source we know that adage and well-known warning of the Poet

> ... The higher they are raised,
> The harder they will fall ...

Finally, is it not true that they who run down a slope receive so great an impetus that, however much they may wish, they cannot stop their forward movement at the bottom, even though they could easily stop it at the beginning?[14]

Riccioli provides more examples, discussing the inability of a clay ball to break an egg or significantly deflect a pan balance when dropped from a low height, discussing the impact of a playing ball that plunges into water from a great height, discussing the greater bounce of a playing ball when dropped from greater heights, and so forth. There are innumerable examples, he says, that make it evident that a heavy body falling from a higher place naturally accrues greater and greater impetus by the end of its motion. However, he says, these examples do not clearly demonstrate the particulars of the motion of a falling body. They do not show whether falling bodies experience a true, uniform growth in velocity as their motion progresses.[15]

And so Riccioli begins his report on his experiments to determine how the velocity of a falling body changes, or as he calls it, "the measuring of the space that any heavy body traverses in natural descent during equal time intervals."[16] He notes that Fr. Niccolò Cabeo had done this at Ferrara in 1634, but only from the relatively short church tower there, and using an uncalibrated pendulum for a timing device. He then launches into a detailed description of the procedure used in his falling body experiments:

> But in 1640 in Bologna I calibrated Pendulums of various lengths using the transit of Fixed Stars through the middle of the heavens. For this [falling body] experiment I have selected the smallest one, whose length measured to the center of the little bob is one and fifteen hundredths of the twelfth part of an old Roman foot, and a single stroke of which [that is, the half period] equals one sixth of a second. . . . As a single Second exactly equals six such strokes, then one single stroke is nearly equal to that time with respect to which the notes of semichromatic music are usually marked, if the Choirmaster directs the voices by the usual measure.
>
> The oscillations or strokes of so short a Pendulum are very fast and frequent, and yet I would accept neither a single counting error nor any confusion or fallacious numbering on account of the eye. Thus our customary method was to count from one to ten using the concise words of the common Italian of Bologna (*Vn, du, tri, quatr, cinq, sei, sett, ott, nov, dies*), repeating the count from one, and noting each decade of pendulum strokes by raising fingers from a clenched hand. If you set this to semichromatic music as I discussed above, and follow the regular musical beat, you will mark time as nearly as possible to the time marked by a single stroke of our Pendulum. We had trained others in this method, especially Frs. Francesco Maria Grimaldi and Giorgio Cassiani, whom I have greatly employed in the experiment I shall now explain.[17]

Having described how to do the measurements, Riccioli feels some obligation to discuss the level of error in the experiment, and the steps his team of priests took to eliminate that error, or at least to keep it under one stroke of a pendulum in time:

> Grimaldi, Cassiani, and I used two Pendulums; Grimaldi and Cassiani stood together in the summit of the Asinelli Tower [in Bologna], I on the pavement of the underlying base or parapet of the tower; each noted separately on a leaf of paper the number of pendulum strokes

that passed while a heavy body was descending from the summit to the pavement. In repeated experiments, the difference between us never reached one whole little stroke. I know that few will find that credible, yet truly I testify it to have been thus, and the aforementioned Jesuit Fathers will attest to this. That is all concerning the Pendulum and the measure of time.[18]

Next Riccioli discusses the place where the experiment was conducted. A variety of buildings and church towers were used, and he notes several in particular. But the place Riccioli's team really liked was Bologna's Asinelli tower, which he notes is 312 feet high, 280 feet from its top to its parapet. The soaring Asinelli was perfect for this sort of experiment, he says, as though it had been constructed for just that purpose. And in a note that reflects a sort of "nerdy" excitement over discovering a cool, ideal opportunity for experimentation that any scientist today will recognize, he gushes, "It is a delight to the eye."[19]

Riccioli includes in his discussion a diagram of the tower (fig. 6.2), labeled to indicate key points on the tower, and noting at what points plumb lines could be lowered for vertical measurements to the base. He notes how the falling balls will drop onto the walled base of the tower, and not into the street. Therefore, "balls can be released from the crown often and frequently without danger to anyone."[20] This diagram serves as a point of illustration for all the experiments; even those that were done using other towers. He includes a summary of his results, which references the tower diagram:

So in May of 1640, and at other times afterward, we determined the height $H\beta$, that from which an eight ounce clay ball, when released, will strike the pavement at precisely five exact strokes of the pendulum described above (that is, 5/6 of a second of time). Through oft-repeated experiments we have discovered this to be 10 Roman feet. Then we determined the height necessary for a ball of the same type and weight to descend in twice as much time, or ten strokes. We discovered this to be 40 feet, which interval KH marks. Ascending further, we determined the appropriate height for thrice as much time, or exactly 15 strokes, which we discovered to be 90 feet, LH. We discovered that for a time of 20 strokes the height MH to be 160 feet, and for 25 strokes the height

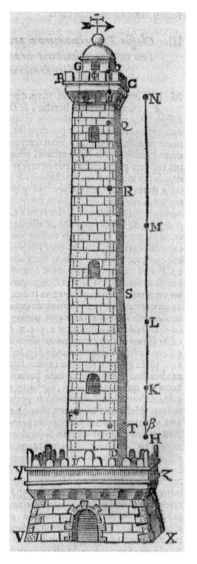

Figure 6.2.   Riccioli's diagram of the Asinelli tower in Bologna, which was the site of many of his falling body experiments (Riccioli 1651, 2:385). The line NH indicates the heights used in the falling body experiments, including those done at other locations. Image credit: ETH-Bibliothek Zürich, Alte und Seltene Drucke.

NH to be 250 feet. Finally, we could not ascend sufficiently high for a ball to require 30 strokes for its descent. So instead we repeatedly released a ball from the crown of the Asinelli tower at G to strike the pavement at I, which is a distance of 280 feet. With me at I and Frs. Grimaldi and Cassiani at G, we consistently counted 26 strokes, as we discovered by comparison of our written notes.

Now let us imagine the intervals marked on the line NH, translated to intervals on the line OT. The distance the ball has travelled at the end of the first five strokes, OC, is 10 feet, and equals βH; the distance the ball has travelled at the end of the second five strokes, OQ, is 40 feet, and equals KH; at the end of the third five, OR, equal to LH, is 90 feet; at the end of the fourth set OS, equal to MH, 160 feet; at the end of the fifth set or 25 strokes the total OT equals the whole NH, 250 feet. Based on the motion prior to T you have some indication of what occurs as the ball continues into the pavement. Therefore the aforementioned ball descends faster and faster the farther it recedes from O and the nearer it approaches to D.[21]

Riccioli notes that in terms of equal time intervals, the distances follow a pattern of 10 feet, 30 feet, 50 feet, 70 feet, 90 feet.

Now Riccioli begins a most interesting discussion in which he compares his results to what Galileo had said in his 1632 *Dialogue*. He writes that he did not understand or recognize how falling heavy bodies increased in velocity in the manner that Galileo described in the *Dialogue*, namely following a pattern of odd numbers begun from unity (1, 3, 5, 7 . . .). This is true, Riccioli says, even though he had the opportunity to make the discovery himself in 1629, when he was with Fr. Daniel Bartolo and Dr. Alphonso Iseo, studying the periods of pendulums, and then again in 1634 when he was working with Fr. Cabeo. In fact, he says, at that time he thought the velocity of a falling body might increase by continually tripling: 1, 3, 9, 27. . . . And, he says, he did not believe Galileo, in part because Galileo's numbers seem plagued by errors and lack of detail:

> [T]he opportunity was granted to me of reading Galileo's dialogues, which the Holy Congregation of the Index had prohibited. I found in the dialogues on page 217 of the Italian or 163 of the Latin[22] the aforementioned growth, discovered by experiment, to be following simple odd numbers from unity, as in 1, 3, 5, 7, 9, 11, etc. Still, I was suspecting something fallacious to lurk in the experiments of Galileo, because in the same dialogue, following page 219 of the Italian, 164 of the Latin,[23] he asserts an iron ball of 100 Roman pounds released from an altitude of 100 cubits reaches the ground in 5 Seconds time. Yet the fact was that my clay ball of 8 ounces was descending from a much greater altitude [280 feet, or 187 cubits], in precisely 26 strokes of my pendulum: 4 and 1/3 Seconds time. I was certain that no perceptible error existed in my counting of time, and certain that the error of Galileo resulted from times not well calibrated against transits of the Fixed stars—error which was then transferred to the intervals traversed in the descent of that ball.[24]

Riccioli figured that these errors then carried over into the rest of Galileo's experiments. Moreover, Riccioli says, he had doubts whether Galileo ever actually dropped a 100 pound ball, which is rather heavy. This is in part because Galileo made no mention of any tower he could have used for such an experiment.

But Riccioli finds that he was wrong, both about his notions of the behavior of falling bodies, and about Galileo:

> And so, full of this suspicion, I began exacting measure of this growth in the Year 1640, as I have said. I hoped to contrive my own idea about this that was nearer to the truth; but rather I have in fact discovered to be true that which Galileo asserted. And indeed as I set forth in the preceding experiment . . . I have acknowledged the growth to follow the proportion of 10, 30, 50, 70, 90 feet, which expressed in smallest numbers is just 1, 3, 5, 7, 9.[25]

Still Riccioli thinks further testing is in order, and so he and Grimaldi ran the experiment for other sets of heights and times. He notes the error level in these experiments—"in the greater distances one or another foot less or more does not introduce a difference of one whole stroke [of a pendulum in time]."[26] Again and again he finds Galileo's "odd number rule" to hold true. The results of all the experiments are presented in a table (see table B.1 and fig. B.2 in appendix B of this volume) "both for the sake of brevity and because fraction numbers are involved."[27]

Now, having learned that he was wrong and Galileo was right, Riccioli set out to spread the word to someone who he thought would like to know—one of Galileo's old followers:

> Therefore Fr. Grimaldi and I went to talk to the distinguished Professor of Mathematics at the Bologna University, Fr. Bonaventure Cavalieri [1598–1647], who was at one time a protégé of Galileo. I told him about the agreement of my experiments with the experiments of Galileo, at least as far as this proportion. Fr. Cavalieri was confined by arthritis and gout to a bed, or to a little chair; he was not able to take part in the experiments. However it was incredible how greatly he was exhilarated because of our testimony.[28]

These experiments turn out to be quite good by modern standards (see appendix B, part 3). This is perhaps to be expected. Riccioli was, after all, ready to spend hours upon hours on the process of merely calibrating a pendulum, and able to talk his colleagues into helping him with such a task. Surely, compared to watching a pendulum swing for hours on end, the falling bodies experiments were far more interesting, and more able to hold everyone's attention. Surely such experiments would receive even greater care.

Riccioli also experimented with falling bodies of differing weights, in essence testing what today would be called the effect of air drag on a falling body. In his report on these experiments, he noted that they could be very misleading unless conducted with great diligence and circumspection. To test the effect of differing weight on falling bodies, he and Grimaldi prepared identical solid clay balls, and then hollow clay balls of identical size, but of half the weight. They also assembled a variety of other balls of differing size and weight. At different times, in front of different observers — men of high character, judgment, and religious integrity, Riccioli notes — they dropped different pairs of balls from the top of the Asinelli tower. Among their observers "three or four Masters of Philosophy or Theology were present, who with Galileo, Cabeo, or Arriaga, had judged that any two heavy bodies, released simultaneously from the same altitude, however great, descend to the ground by the same physical moment of time."[29] These Masters promptly set aside that opinion upon seeing the experiments, because the solid balls consistently struck the ground first. The solid clay balls were consistently faster to the ground than the hollow clay balls, by more than half a second (the total fall time was around 4 seconds), and the gap between the two was evident by the time they were halfway down the tower.

Riccioli presents a table of data listing different pairs of balls, along with which ball in each pair struck the ground first in the experiments and by how much distance (see table B.2 and fig. B.3 in appendix B in this volume). From this table he draws six conclusions regarding pairs of spheres simultaneously released from equal heights. First is that if the two spheres are equally heavy and have the same density, they will reach the ground in equal times. If one of those were somehow expanded or compressed so as to change its density but not its weight, then the smaller of the pair would reach the ground first, on account of reduced air drag.[30] Second is that if the two spheres have equal density but unequal weight, the one which is heavier reaches the ground first. This is true whether they are of equal size (because one of them is hollow), as in experiment 1 in the table (repeated twelve times) and experiment 2, or of differing size, as in experiments 7, 8, 9, 13, 14, 15, 16, and 21. Third is that if the two spheres are equally heavy, but not equally dense, the one that is denser reaches the ground first, as seen in experiment 12, owing to reduced drag—Riccioli notes that the denser sphere is "sharper of shape or angle of contact, owing to the small-ness of the sphere."[31] Fourth is that if one of the two spheres is both heav-ier and denser than the other, it reaches the ground first, whether it is larger in size than the other, as in experiment 10, or equal to the other, as in experiments 3, 4, 5, 6, or smaller than the other, as in experiment 11. Fifth is that if one of the two spheres is denser, but not heavier, then no clear rule applies. The more dense sphere may reach the ground first (ex-periments 12, 17, and 18), last (experiments 19 and 20), or in equal time (comparing experiment 8 with 18). Sixth is that the balls that are both lighter and less dense—in other words, that have both lesser individual gravity and lesser specific gravity—always reach the ground last. In this sense "heavier bodies fall faster, at least in our air."[32]

Riccioli notes that sometimes balls of differing weights can reach the ground at nearly the same time—the smaller size of a ball can allow it to punch through the air more readily and reach the ground in about the same time as a heavier ball. He notes,

> Experiments 19 and 20 indeed disturb me. In these a little lead ball of one ounce, acquired more (or at least nearly as much) velocity on account of smallness, than a heavier clay acquired on account of far greater weight. The more perfect roundness of the lead than of the clay contributed to this.[33]

Recall from chapter 3 that Tycho Brahe believed that air was too "fluid" and "tenuous" to affect in any significant way the motion of a heavy body traveling through it. Riccioli's experiments suggest this not to be the case, which would have serious implications for the physics of impetus to which Tycho and others subscribed. Riccioli's six conclusions regarding the effects of air on falling bodies with differing characteristics conform with modern understanding of the physics of falling bodies subject to air drag: drag effects will be lesser in smaller, denser bodies (see appendix B, part 3).

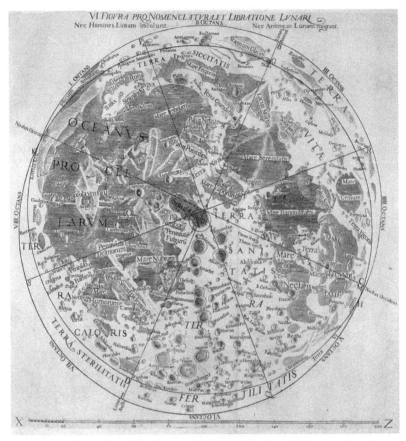

Figure 6.3.   Map of the Moon from the *New Almagest* (Riccioli 1651, 1:204–5). Image courtesy History of Science Collections, University of Oklahoma Libraries.

Riccioli's falling bodies experiments tell us much about Riccioli. He is thorough. He provides a full description of his experimental procedure. His data, which include estimates of uncertainty or error for his measurements, are of sufficient quality to determine the acceleration due to gravity (g) to an impressive accuracy (appendix B, part 3). He undertakes an experiment expecting to disprove Galileo's ideas (distrusting Galileo in part because Galileo does not report the sorts of experimental details that he would), yet when his results instead confirm Galileo's ideas, he makes a point of promptly sharing the exciting news with an interested colleague who worked with Galileo, and he later publishes the results.

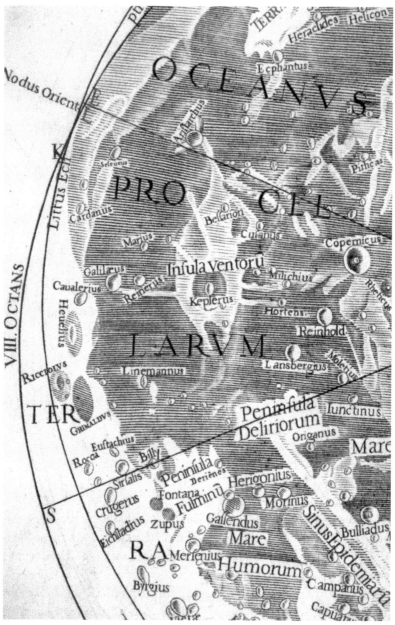

Figure 6.4. Detail from the *New Almagest* map of the Moon (Riccioli 1651, 1:204–5). Image courtesy History of Science Collections, University of Oklahoma Libraries.

The falling body experiments do not stand alone in illustrating Riccioli's thoroughness. The *New Almagest* is filled with extensive reports on different experiments, and with tables of data from real experiments, reported whether that data fit a particular model or not. It reflects close, careful work beyond just the timing of falling bodies—such as the work necessary to determine that the period of a pendulum is independent of its amplitude only for small-amplitude oscillations, whereas larger oscillations have a longer period. Riccioli illustrated the reliability of his work by providing descriptions of how it was carried out, so those who wished to reproduce his experiments could do so.[34] The *New Almagest* map of the Moon also reflects thoroughness and attention to detail (figs. 6.3 and 6.4). Riccioli appears to be not just a scientist, as Grant has said, but something of a model scientist who designed and conducted experiments of great precision, recorded the procedure of those experiments so others could reproduce them, and accepted and reported the results of those experiments, even when they contradicted his ideas and supported Galileo's.

Let us keep the falling body experiments in mind as we learn more about Riccioli. Obviously we cannot be certain that, just because Riccioli was accurate, thorough, and objective in this work, he will necessarily be accurate, thorough, and objective in all his work. However, we have seen what he can do. We ought not dismiss him like (we shall see) others have often done.

7   126 Arguments

Giovanni Battista Riccioli's *New Almagest* contains information on a great range of topics in astronomy, from the positions of stars to the telescopic appearance of planets. But its centerpiece is a full book, book 9, dedicated to an analysis of the debate over whether or not the Earth moves as hypothesized by Copernicus: annually around the Sun, with diurnal rotation around its own axis. Edward Grant has described this as "probably the lengthiest, most penetrating, and authoritative analysis made by any author of the sixteenth and seventeenth centuries."[1] The debate Riccioli analyzes is, of course, between the heliocentric hypothesis of Copernicus and the hybrid geocentric hypothesis of Tycho Brahe[2]—the old, purely geocentric hypothesis of Ptolemy having been cast aside on account of telescopic observations such as the phases of Venus and Mercury. As we saw in chapter 1, the frontispiece of the *New Almagest* reflects all this (fig. 1.1). And so it is the Tychonic and Copernican hypotheses Riccioli has in mind when in book 9 he presents 126 arguments from the debate over the Copernican motion of the Earth—49 arguments for Earth's motion, 77 against—along with responses to each argument from the other side in the debate.

History has not been kind to Riccioli's discussion of pro- and anti-Copernican arguments. Across a period of two centuries, source after source has offered negative assessments of the *New Almagest* analysis:

[Riccioli was] an enemy to the Copernican system, and has the discredit of having measured the evidence for and against that system, not by the weight but by the number of the arguments.[3]

Father Riccioli put forward a series of arguments in contradiction of the Earth's movement. These arguments, seventy-seven in number, were marvellously absurd. "Would the birds," said Father Riccioli, for example, "dare to rise in the air if they saw the earth passing away from beneath them?" From such a specimen we may judge of the rest of the egregious structure.[4]

Thorough and up-to-date though it was, Riccioli's treatment of the cosmological controversy was a sterile exercise. Astronomical issues could no longer be settled by a preponderance of scientific and scriptural authority or by any number of decrees from Rome.[5]

In Riccioli's ultimate acceptance of the immobility of the earth, biblical and theological arguments proved decisive.[6]

Riccioli had no real arguments to support the geocentric system other than the Bible and the authority of the Church.[7]

Riccioli was a serious astronomer and knew that Ptolemy's universe could no longer be upheld, but his religious beliefs forced him to argue against the Copernican hypothesis. . . . In the [New Almagest], he produced forty-nine arguments that were in favor of heliocentrism, and seventy-seven that were against, and thus the weight of the argument favored an Earth-centered cosmology![8]

This negative assessment of Riccioli's analysis is unfortunate, and perhaps Riccioli himself bears some blame for it. The New Almagest can seem almost like an endless procession of oversized pages packed with dense text, often in the form of lengthy sentences (some exceeding 150 words), supplemented by myriad internal cross-references. Riccioli is obviously writing for the reader who is as dedicated and thorough as himself, and most readers probably have not been so. However, the reader who does tackle the New Almagest will find the above sources to be most inaccurate in their assessment of its analysis.[9] Indeed, he or she will find these sources to be correct only in that Riccioli analyzed 126 arguments in the debate over Copernicanism, and that (as his frontispiece with its balance tipping in favor of a hybrid geocentric world system shows) in the end he believed the arguments to weigh against the Copernican system.

The first thing the reader of the *New Almagest* will likely note is that Riccioli definitely is not presenting 126 arguments that he considers to be persuasive. No, to Riccioli, there are valid responses (valid counter-arguments) to all of the 49 pro-Copernican arguments and to most of the 77 anti-Copernican arguments. Consider the "birds" argument offered above as the specimen by which the rest of the "egregious structure" of Riccioli's analysis may supposedly be judged. Riccioli indeed does present this argument—it is his anti-Copernican argument number 27. But he also notes that this argument can be answered by the Copernicans, "because in the Copernican hypothesis birds, ships, etc. have not only their own motion, but also a common motion by which they move equally with the Earth."[10]

Moreover, Riccioli is not presenting 126 arguments that he feels have worth. Among the 126 are quite a few he rejects as being based on ignorance. The reader may be surprised to see that Riccioli rejects as many or more anti-Copernican arguments for such reasons as he does pro-Copernican. For example, he presents the anti-Copernican argument (number 60) that says the changes in length of days and nights cannot be explained in the Copernican hypothesis—an argument Ingoli used in his essay (see chapter 5)—but then notes that this argument simply arises out of ignorance of that hypothesis, and he directs the reader to a lengthy discussion of the matter elsewhere in the *New Almagest*.[11] The very next anti-Copernican argument (number 61) that he presents claims that lunar eclipses, with the Moon and Sun opposite each other in the zodiac, might not occur were Copernicus correct. Riccioli notes that in the Copernican hypothesis the Earth carries the orbiting Moon with it around the Sun, so the Earth can still be interposed between Sun and Moon for an eclipse, as those who understand the mechanism of eclipses will realize.[12]

Some of the 126 Riccioli rejects as being simply inane. Pro-Copernican argument number 15 is particularly noteworthy in this regard. Riccioli dutifully presents the argument:

If the Fixed Stars were moved by diurnal motion rather than Earth . . . some might trace out the largest circles (those at the Equator); others the smallest (those near the poles). Thus the former would be swiftest, the latter slowest. But this is absurd. Therefore let Earth be moved by diurnal motion, rather than the Fixed Stars.[13]

Riccioli provides the response to this argument, which is, in essence, that such is motion on any rotating sphere, be it the sphere of stars or the sphere of the Earth, and if anything, such varying speeds are no less absurd on Earth than in the heavens.[14] But before providing this response Riccioli feels compelled to offer some commentary on this argument that Galileo had included in his *Dialogue:*[15]

Actually I might be ashamed of having mentioned this argument, except that otherwise all the shame will pour onto Galileo, who (by all means, ready for a flogging by a Neophyte of Astronomy) seriously promotes it.[16]

He also rejects as inane another argument from the *Dialogue* (pro-Copernican argument number 12), which says that the progression of the periods of celestial bodies is logical in the Copernican system (where the periods of planets increase from swift Mercury near the Sun, out to slow Saturn, and on to the unmoving stars), whereas it is not logical in a geocentric system. This is because, as Galileo writes, if Earth is to be motionless, then,

it is necessary, after passing from the brief period of the moon . . . to that of Mars in two years, and the greater one of Jupiter in twelve, and from this to the still larger one of Saturn, whose period is thirty years—it is necessary, I say, to pass on beyond to an another incomparably larger sphere, and make this one finish an entire revolution in twenty-four hours.[17]

Riccioli points out that this argument does not hold, because the periods of Mars, Jupiter, and Saturn are measured with reference to the stars, whereas the twenty-four hour period is measured with reference to the horizon. Mars, Jupiter, and Saturn each also rise and set every twenty-four hours. Riccioli calls this an ugly use of fallacy and equivocation that, again, an astronomical neophyte would recognize.[18]

But Riccioli considers most of the 126 arguments to be neither inane nor matters of ignorance. Most of the arguments he presents, from either side of the debate, are reasonable, but are nonetheless answerable by the opposing side in the debate. Therefore they are not decisive. Previously we saw the birds argument, which Riccioli noted is answerable by common motion (in fact, Riccioli presents quite a number of anti-Copernican arguments that he notes can be answered by common motion). In chapter 1 we encountered the argument about centers, and the argument about the location of hell, both of which were answerable. A look at a few more examples of what Riccioli might call reasonable arguments with reasonable answers is in order.

First, let us consider pro-Copernican argument number 6:

Diurnal motion should be attributed to a body that is definitely understood to be mobile, rather than to one concerning whose mobility we are not certain. We are certain concerning the mobility of the Earth, because we are certain that it is finite. We are uncertain concerning the mobility of the highest heaven [the stars] because we are uncertain whether it be finite or infinite. For, if it is infinite, either it is not movable by diurnal revolution, or, at least it is controversial among Physicists as to whether it is movable.[19]

This is a fine argument, but Riccioli notes it can be answered with a fine response. That response is that either the fixed stars move, or we move (with the Earth). The same sensory evidence, physics experiments, calculations, and so forth, that indicate that Earth is finite—for we do not experience Earth's size and shape directly—also indicate that the stars move.[20] (As we shall see, Riccioli produced strong physical arguments for the motion to belong to the stars).

Or, consider anti-Copernican argument number 39, which is that, according to Aristotle, heavy bodies tend toward and light bodies recede from the center of the universe, not the center of the Earth. Riccioli says the Copernican response to this is that heavy bodies carried by the Earth tend towards the center of the Earth (that is, toward the center of the heaviest body), while light bodies tend toward the circumference of the elemental system, which Aristotle has not proven to be concentric to the universe. Riccioli mentions both Galileo and Kepler in connection with this response. He criticizes Kepler, but he approves of Galileo's response.[21] In the *Dialogue*, Galileo states that he does not believe that there is some center point in the universe that heavy things gravitate toward, but rather that

> these materials, cooperating naturally toward a juncture, would give rise to a common center, this being that around which parts of equal moments are arranged. From this I suppose that the large aggregation of heavy bodies being transferred to any place, the particles which were separated from the whole would follow.[22]

The idea that the Earth lies at the center of a spherical elemental system that circles the sun as a whole, and within which Aristotelian physics and the Aristotelian elements hold sway (fig. 7.1), plays a prominent role in the Copernican answers to a number of the anti-Copernican arguments that Riccioli discusses.[23] It also arises a number of times among both the anti-Copernican arguments and the responses to pro-Copernican arguments. While this concept is not unreasonable, Riccioli notes that an orbiting elemental system is awkward in how it puts the elemental sphere in motion. He notes that it is inelegant in how it complicates the natural motions of Earthly bodies, so that a heavy body has a natural rectilinear motion toward Earth's center plus two natural curvilinear motions (diurnal and annual). And he notes that it is ad hoc in postulating the motion of the heavy Earth around the Sun without a physics to explain how such a heavy body moves (Isaac Newton's physics being still a generation in the future when the *New Almagest* was published).[24]

Among the reasonable arguments that Riccioli discusses from both sides of the debate are some that are especially interesting, even if Riccioli did not consider them to be decisive. These are worth some further discussion, for they illustrate some of the insightful thinking present on both sides of the debate. One very interesting line of argument that Riccioli discusses (it is involved in pro-Copernican argument number 49 and anti-Copernican argument number 1) is one about falling bodies that appears in Galileo's *Dialogue*. Galileo had proposed (and supported with geometrical constructions) that the apparent linear acceleration of a stone falling from a tower might be the result of two uniform circular motions: the

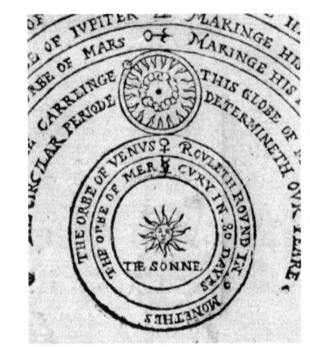

Figure 7.1.
Detail from Digges's "Perfit Description" illustration, showing the "sphere of the elements" traveling around the Sun with the Earth. Elemental fire is just within the sphere of the Moon, with air below it, and the Earth itself at the center. Within this sphere is the realm of Aristotelian physics and elements. Heavy objects move toward its center, while fire rises toward its circumference. Image courtesy History of Science Collections, University of Oklahoma Libraries.

diurnal rotation of Earth, and a second uniform circular motion belonging to the stone. This second circular motion had the same circumferential speed as the diurnal motion at the top of the tower, yet was centered on a point located halfway between the Earth's center and the top of the tower (see fig. 7.2).[25] Thus, Galileo says,

> [T]he true and real motion of the stone is never accelerated at all, but is always equable and uniform. . . . So we need not look for any other causes of acceleration or any other motions, for the moving body, whether remaining on the tower or falling, moves always in the same manner; that is, circularly, with the same rapidity, and with the same uniformity.[26]

Galileo goes on to say that the movement of a falling body is either exactly this, or very near to it, and that

> straight motion goes entirely out the window and nature never makes any use of it at all.[27]

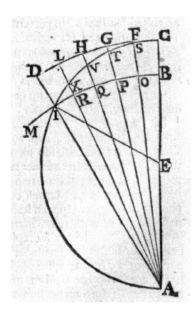

Figure 7.2. Diagram from the *New Almagest* (Riccioli 1651, 2:399) showing Galileo's hypothesis that the motion of a heavy ball falling from a high tower might be the result of two natural circular motions. Point A is the center of the Earth. Curve BM shows the circular path of the base of the tower on Earth's surface as Earth rotates counter-clockwise. Curve CD is the circular path of the top of the tower. Curve CIA is the proposed natural circular path of the ball when released. It is centered on E, a point located halfway between the top of the tower and the center of the Earth A. The speed of the ball along CIA is the same as it was along CD. The apparently linear fall of the ball, by which it appears to move away from the top of the tower and toward the center of the Earth, is explained by the increasing gap between curve CD, which the tower top follows, and curve CIA, which the ball follows. This gap is marked by intervals FS, GT, HV, LX, DI. Riccioli noted that the downward acceleration that would result were this hypothesis valid is much smaller than the actual measured acceleration of a falling body, and that this theory does not, for example, explain the fall of a body at the poles, where there is no diurnal motion. Image credit: ETH-Bibliothek Zürich, Alte und Seltene Drucke.

Thus here could be a new physics to explain motion in the Copernican system. In this physics, all natural motion, including that of heavy objects such as stones, is circular. The Copernican motion of Earth is thus natural. Rectilinear motion does not exist, and what appears to be rectilinear motion, like the falling stone, is the result of a combination of natural circular motions. If Galileo is right, then the Copernican hypothesis is far stronger. As Owen Gingerich has emphasized in his writings, credible scientific explanations hang together in a tapestry of coherency that supports observations.[28] This would be a coherent explanation for all natural motions in the Copernican system.

However, a rigorous analysis of Galileo's hypothesis leads to experimentally testable predictions regarding the rate of acceleration of a falling

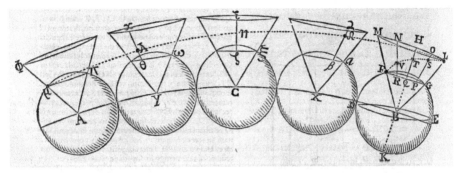

Figure 7.3.   Diagrams from the *New Almagest* (Riccioli 1651, 2:404) showing the motions of falling bodies on a rotating, Sun-orbiting Earth. Image credit: ETH-Bibliothek Zürich, Alte und Seltene Drucke.

body. As we saw in chapter 6, Riccioli devised precise experiments to measure this rate of acceleration. This rate turns out to be far greater than the rate expected according to Galileo's hypothesis that all natural motion is circular.[29] Moreover, except at the equator, the hypothesis brings forth a variety of peculiar motions and impetus effects, as the falling body's motion is quite complex when not on the equator (fig. 7.3), and at the pole there would be no circular diurnal motion at all. Thus this most interesting argument of Galileo's is answered by experiment and by further reflection on what the argument implies.[30]

The pro-Copernican argument on sunspots (number 43) is another interesting example. The argument is that the apparent paths of sunspots across the face of the Sun are explained by simpler and fewer motions if the Earth, rather than the Sun, moves with annual motion. Riccioli notes that anti-Copernicans can answer that in either the geocentric or the heliocentric hypotheses only three real motions are necessary to explain the sunspot motion. In the heliocentric hypothesis these are the annual motion of Earth, the diurnal rotation of Earth, and an approximately monthly rotation of the Sun. In a geocentric hypothesis the three motions are an orbital motion of the Sun around the Earth that is slightly slower than the motion of the Fixed stars (and thus accounts for both the diurnal and annual motions of the Sun—see chapter 2), an annual gyration of the poles of the Sun, and an approximately monthly rotation of the Sun. And Riccioli notes that, while there may be a parity between the two hypotheses regarding the number of motions, anti-Copernicans will argue that the heliocentric system is the inferior arrangement. This is because, if it is the Earth that moves, the motions are divided between terrestrial and celestial, and the two terrestrial motions are inferred, not directly observed. On the other hand, if it is the Sun that moves, all three motions are celestial, and all three are observed.[31]

A third example of some of the particularly interesting arguments from the 126 is pro-Copernican argument number 48, regarding the tides of the sea. This is Galileo's famous argument from the *Dialogue*.[32] In presenting the argument, Riccioli relates Galileo's theory: tides "are explained by no other more suitable and evident manner, or at least not adduced by a more probable cause, than through unequal motion of the Earth arising by reason of the combined diurnal and annual motion."[33] The anti-Copernican response is that the effect Galileo describes (the changing speed of Earth's surface resulting from its combined annual and diurnal motions) would probably be too small to cause the tides. And even were it not too small, it would fail to explain the observed variations in the tides (the papal commission charged with reporting on the *Dialogue* in 1632 particularly pointed out the inability of Galileo's theory to account for the observed twice-daily high and low tides, when the varying speed of Earth's surface would suggest a once-daily high and low tide).[34] It also overturns the best nautical knowledge, which connects the tides with the phases of the Moon. Although he can't resist grousing about "the unrestrained boasting of Galileo in this argument," Riccioli grants that, "thus far no opinion that satisfies the contemplating intellect has sprung forth concerning the cause of the tides of the Sea that eliminates the many difficulties, and accounts for the many differences of the tides." But he insists that, nevertheless, "there are some that are more probable than the opinion of Galileo."[35] He refers the reader to elsewhere in the *New Almagest* for a very extensive discussion of the tides. He also mentions that, if the tides are to be explained by the combinations of an orbital motion of the Earth combined with a diurnal rotation, it would be better to suppose that the Earth circles the Moon, an idea developed in jest by Giovanni Battista Baliani (1582–1666).[36]

Elsewhere Riccioli addresses the interesting question of whether the same combination of Earth's supposed annual and diurnal motions that Galileo proposed as the cause of the tides might affect the period of a pendulum over the course of a day—this he brings up in relation to a pro-Copernican argument (number 42) that says Marin Mersenne (1588–1648) claimed such an effect could be observed. Riccioli and Grimaldi conducted experiments to look for such an effect, yet

> neither in the morning, nor in the evening, nor near Midday nor Midnight (which might be better, because then the difference that comes from the combination of diurnal and annual motions might be more evident than in the morning or the evening), has any certain and sensible difference appeared between the equivalents of one hour noted from the transits of stars, and noted from vibrations of the pendulum.[37]

He even taps this same concept as an answer to the pro-Copernican argument (number 32) that heliocentrism eliminates the complex, looping, "retrograde" motion of the planets (see chapter 3), rendering it mere appearance. He notes that the combination of Earth's motions effectively introduces a daily retrograde motion for every object on Earth, a motion that Riccioli says is "unnecessary and lacking any foundation obtained from the senses, which can perceive no motion of the Earth at all, much less any variation in that motion."[38]

A last example of some particularly interesting arguments—or, more correctly, interesting responses to arguments—are the last two that Riccioli presents. These are anti-Copernican arguments numbers 76 and 77, which pertain to a lack of an absolute point of reference in the Copernican system. The first of these is that since the centers of the Earth and the universe are separated by the radius of Earth's orbit, it is uncertain from which center the distance to the stars should be measured. To this the Copernican response is: "This measure might be estimated from both centers, although by different ways."[39] The second is that the Copernican hypothesis grants license to place any planet at the center of the Universe, to which the response is that any hypothesis must explain the celestial phenomena, and none that do are more suitable than that of Copernicus.[40] The lack of an absolute reference point is not, for Riccioli, a decisive issue.

As interesting as these arguments are, to Riccioli's mind they all had something very important in common with the overwhelming majority of the 126 arguments: they could be answered. To Riccioli, all of the arguments we have discussed—be they weak arguments based in ignorance, inane arguments that he thought really ought not be mentioned, or just solid, reasonable arguments—were ultimately not decisive in the debate over the Copernican hypothesis. The decisive arguments, those that caused the frontispiece scales to tip, were the ones that could *not* be answered. There were only a handful of these. All of them were anti-Copernican. All were rooted in the ideas of Tycho Brahe.

8   An Angel and a Cannon

We have seen how Giovanni Battista Riccioli said the bulk of the 126 arguments for and against the Copernican hypothesis that he analyzed ultimately were not decisive. The decisive arguments were few, and they were anti-Copernican. They clustered around two phenomena, both of which Tycho Brahe had brought up in his anti-Copernican arguments. One of these two phenomena was the trajectories of falling or thrown heavy bodies: cannon balls dropped from towers, or launched from a gun in different directions.

Recall from chapter 3 that Tycho argued that Earth's motion should be detectable through physical experiments—that we should see some evidence of Earth's motion in phenomena visible here on Earth. He discussed the subjects of a lead ball dropped from a tower, of identical cannons launching identical balls to the east and to the west from a given point, and of cannons at the equator and at the poles launching identical balls in various directions.

The Copernican answer to such arguments was common motion: the surface of the Earth and everything on it moves with a common motion; the net result is that all effects are the same as if there were no motion at all. For this reason Galileo had argued in his *Dialogue* that physical experiments would not reveal evidence of Earth's motion, as we saw in chapter 3. Citing the example of how the behavior of all sorts of things in the cabin of a ship would occur the same way whether the ship was at rest or moving smoothly, and how "the cause of all these correspondences of effects is the fact that the ship's motion is common to all the things contained in it, and to the air also,"[1] Galileo declares (through the character of Sagredo) that "I am satisfied so far, and convinced of the worthlessness of all experiments brought forth to prove the negative rather than the affirmative side as to the rotation of the earth."[2]

Sagredo should not have been satisfied. Within the *Dialogue* itself is a discussion of an experiment that is not refuted by the common motion argument, namely, the firing of a cannon at targets to its east and west.[3] Galileo writes concerning targets that are close enough to the cannon that the balls launched toward them are not significantly deflected downward by gravity. Thus they can be regarded as traveling along a line tangent to the Earth's surface—a line that is fixed relative to the stars. Stated another way, if a certain star is located behind each target at the moment of firing, the cannon balls travel toward those stars, along the tangent line. The rotation of the Earth, meanwhile, causes the targets to move relative to the stars (fig. 8.1). Earth's rotation causes the western target to rise as the ball travels to it, so the star behind it sets. The west-moving ball, which is traveling toward the star, thus will strike below its target. Likewise, Earth's rotation causes the eastern target to drop as the ball travels to it, so the star behind it rises, and so the east-moving ball will strike above its target.

This argument involves a rotational motion. Earth is a spherical, rotating body. It is what physicists today would call a rotating frame of reference. It is not a flat translating (moving in a straight line) frame of reference, like the cabin of Galileo's ship. Thus the ship analogy, which convinces Sagredo of the worthlessness of all such experiments, does not apply. Galileo argues that this effect is small, if it is real (through the character of Salviati, Galileo leaves open the possibility that the trajectories of projectiles are locked to the line between their projector and target).[4] He calculates the deflection to be roughly an inch for targets 250 yards away, and thus to be undetectable compared to the accuracy of any cannon at that range.[5]

Riccioli presents this argument in the *New Almagest* (anti-Copernican argument number 18), along with Galileo's response to it.[6] This argument is important. Despite Galileo's assertion of "the worthlessness of all experiments brought forth to prove the negative rather than the affirmative side as to the rotation of the earth," in fact this experiment should

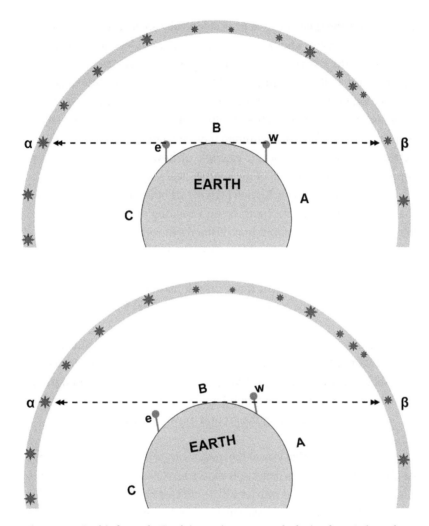

Figure 8.1.   In this figure the Earth is rotating counter-clockwise, from A through B toward C. Top: Cannons at B launch balls toward targets to the east (e) and west (w). Those balls travel along a line tangent to Earth's surface toward stars α and β. Bottom: While the balls are in flight, the Earth rotates, causing target e to drop below the tangent line, and target w to rise above it (seen from Earth, star α rises above target e, and star β sets below target w). The result is that the east-moving ball strikes above its target, while the west-moving ball strikes below its target.

work (unless the trajectories of projectiles really are locked to the line between their projector and target). Perhaps a cannon does not have the accuracy to do the experiment, but granted a projectile launcher of sufficient
accuracy, the deflection effect *should* be detectable, and thus so should
the rotation of the Earth. Tycho Brahe had merely provided some general
ideas about how Earth's rotation might be detected in projectiles. Galileo,
despite Sagredo's claim that common motion rendered Earth's motion undetectable, gives a very specific example of how to detect that motion,
complete with a calculation of the approximate magnitude of the effect.

Among the arguments Riccioli presents are others that deal with effects
that should arise from the Earth being a rotating frame of reference. He
cites Tycho's argument about how, if the Earth rotates, balls fired from a
cannon at one of the poles would have the same trajectory regardless of the
direction in which they were fired, while balls fired from a cannon at the
equator would have different trajectories if fired toward the east or west
than if fired toward the north or south. Indeed, writes Riccioli,

> if a ball is fired along a Meridian toward the pole (rather than toward
> the East or West), diurnal motion will cause the ball to be carried off
> [that is, its trajectory is deflected], all things being equal: for on paral-
> lels nearer the poles, the ground moves more slowly, whereas on par-
> allels nearer the equator, the ground moves more rapidly.[7]

Riccioli declares that even if the trajectories of the cannon balls cannot
be observed directly, this deflection should reveal itself when a ball strikes
a target, since its angle of impact will be affected.[8] He even illustrates this
effect (fig. 8.2). His anti-Copernican arguments numbers 17 and 19 derive
from these ideas—arguments to which, he notes, the Copernicans have no
solid answers.[9] He reports that skilled artillerymen are accurate enough
with their shots to be able to place a ball directly into the mouth of an
enemy's cannon, so the difference in shots east/west versus shots north/
south should have been detected, if it existed.[10]

Riccioli presents another argument based on Earth's rotation. In his
anti-Copernican argument number 10, he imagines a massive sphere attached to a chain and dropped from a point at a great height above the
Earth's surface:

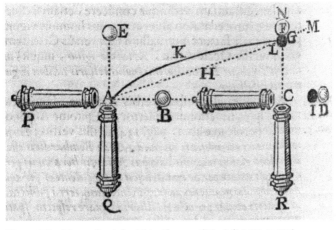

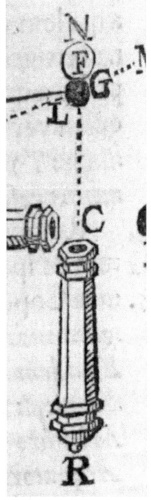

Figure 8.2. Figure from the *New Almagest* (Riccioli 1651, 2:426), showing the trajectories of a cannon fired to the north versus fired to the east—part of Riccioli's Coriolis effect argument against the Earth's diurnal rotation. Riccioli writes that if the cannon is fired eastward at a target at B, then as the ball is in flight, the Earth's (supposed) diurnal rotation carries the mouth of the cannon from A to C, and carries the target from B to D, so the ball travels from A to I, where it hits the target at D. If the cannon is aimed northward and fired at a target at E, then as the ball is in flight, the target moves from E to N. However, he says, the ball travels along the curve AKF, not the straight line AHF. This happens because the diurnal motion is faster at the beginning of the ball's flight. The ball will not strike the target at N squarely, but will graze it or miss it. However, he says, if another target were positioned east of N, such as at G, the ball will squarely strike it, even though the cannon is not aimed at it. Riccioli believed a skilled artilleryman could place a shot precisely into the mouth of an enemy's cannon, so the difference in shots east/west versus shots north/south should have been detected, if it existed (Riccioli 1651, 2:426–27).

Riccioli credits the physics of this analysis to Grimaldi. Grimaldi's work is consistent with an analysis using modern physical principles. Were all the ground moving eastward at a constant rate, the ball would follow a straight path AHF from the cannon at A to the end of the ball's flight at F when it strikes the target at N. However, since the ground speed progressively decreases as the ball heads north, the freely flying ball (whose speed does not change) progressively outpaces the ground, with the result that its trajectory bends progressively to the east relative to the ground, much as in this figure. Both ball and target are moving east, but since the ball outpaces the target, it strikes to the east of the target. The cannon travels with the ground and the target, so as seen from the cannon (see detail), the ball, expected to travel in a straight line from the mouth of the cannon at C to the end of its travel at F, is in fact seen to veer slightly right, into G (see Graney 2011c). Images credit: ETH-Bibliothek Zürich, Alte und Seltene Drucke.

If an Angel were to let fall a metal sphere of great weight hung to a chain, while holding the other end of the chain immobile, that chain by the force of the sphere might be extended to its full length perpendicularly toward the Earth; but following the Copernicans, it ought to curve obliquely toward the East.[11]

In other words, the sphere should drop straight to the ground if the Earth is immobile, but will be deflected eastward if the Earth has a diurnal rotation. Here Riccioli is presenting an argument that looks something like Tycho's argument about a ball falling from a tower (recall from chapter 3). However, whereas Tycho expects the ball to deflect to the west, this argument says it should deflect to the east. The top of the tower is further from the center of the Earth than the bottom, and if the Earth is rotating toward the east as Copernicus says, the top must be moving faster than the bottom. The ball is traveling east at the speed of the top of the tower, and once released it still retains that speed as it falls (Riccioli apparently had a better grasp of Buridan's physics of impetus—recall from chapter 2— than did Tycho, or he had read Galileo's new writings on physics). Thus if the Earth is rotating, the ball should outrun the base of the tower as it drops, and strike the ground somewhat toward the east of the spot where it would strike were the Earth not moving.

The modern reader who has studied physics or meteorology may recognize in all these arguments, and especially in Riccioli's cannon illustration, the *Coriolis effect*—an illusory or fictitious "force" on projectiles and falling bodies, which arises on account of the Earth being a rotating frame of reference. In fact, projectiles are deflected much as Riccioli described in his cannon discussion. This effect is indeed known to gunners, having become apparent in the nineteenth century as guns increased in range and accuracy. It can be significant over long ranges.[12] Riccioli recognized that this effect would be more noticeable for long-range projectiles, stating that it will not be insensible "when the motion [of the projectiles launched in the different directions] is violent to sufficient degree."[13]

While the modern reader can probably imagine many problems in trying to detect the rotation of the Earth by measuring the Coriolis deflections of artillery projectiles—it would be all but impossible, for example, to fire truly identical cannon balls at identical angles and velocities and so forth—

measuring deflections in falling heavy objects is a different matter. This would seem to be simply a matter of carefully measuring with a plumb line, and carefully releasing a heavy ball from rest. It is not difficult to make a rough calculation of the expected eastward deviation of a body falling from a tower at the equator.[14] The expected deflection for a fall from a height such as Riccioli's Asinelli tower is roughly an inch: Giovanni Alfonso Borelli (1608–1679) published such a calculation in 1667, as part of a debate prompted by Riccioli's work.[15] Thus, this eastward deflection is small, but would seem to be detectable, especially if the influence of the air is negligible, as would be expected when the falling body is dense and heavy.

Two people who thought that this eastward deflection of a falling body would be detectable, were the experiment to be carried out, were Isaac Newton (1643–1727) and Robert Hooke (1635–1703) of England. Hooke was familiar with Riccioli's analysis of pro- and anti-Copernican arguments. In his 1674 essay, *An Attempt to Prove the Motion of the Earth by Observations*, Hooke wrote that only one anti-Copernican argument would carry any weight:

> The Inquisitive Jesuit Riccioli has taken great pains by 77 Arguments to overthrow the Copernican Hypothesis, and is therein so earnest and zealous, that though otherwise a very learned man and good Astronomer, he seems to believe his own Arguments; but all his other 76 Arguments might have been spared as to most men, if upon making observations as I have done, he could have proved there had been no sensible Parallax this way discoverable, as I believe this one Discovery will answer them, and 77 more, if so many can be thought of and produced against it.[16]

Annual parallax would not be successfully detected for another century and a half, but Hooke would soon be attempting to prove the motion of the Earth by another means—Riccioli's eastward deflection of a falling body. A letter that Newton wrote to Hooke in November 1679 brought this argument to Hooke's attention:

> I shall communicate to you a fancy of my own about discovering the earth's diurnal motion. In order thereto I will consider the earth's

diurnal motion alone, without the annual, that having little influence on the experiment I shall here propound. Suppose then BDG [see fig. 8.3] represents the globe of the earth carried round once a day about its center C from west to east according to the order of the letters BDG; and let A be a heavy body suspended in the air, and moving round with the earth so as perpetually to hang over the same point thereof B. Then imagine this body [A] let fall, and its gravity will give it a new motion towards the center of the earth without diminishing the old one from west to east. Whence the motion of this body from west to east, by reason that before it fell it was more distant from the center of the earth than the parts of the earth at which it arrives in its fall, will be greater than the motion from west to east of the parts of the earth at which the body arrives in its fall; and therefore it will not descend the perpendicular AC, but outrunning the parts of the earth will shoot forward to the east side of the perpendicular describing in its fall a spiral line ADEC, quite contrary to the opinion of the vulgar who think that, if the earth moved, heavy bodies in falling would be outrun by its parts and fall on the west side of the perpendicular. The advance of the body from the perpendicular eastward will in a descent of but 20 or 30 yards be very small, and yet I am apt to think it may be enough to determine the matter of fact.[17]

Newton's idea was discussed at a meeting of the Royal Society on 4 December 1679. According to the minutes of this meeting,

Mr. Hooke produced and read a letter of Mr. Newton to himself, dated 28th November, 1679 . . . suggesting an experiment, whereby to try, whether the earth moves with a diurnal motion or not, viz. by the falling of a body from a considerable hight, which, he alledged, must fall to the eastward of the perpendicular, if the earth moved.

This proposal of Mr. Newton was highly approved of by the Society; and it was desired, that it might be tried as soon as could be with convenience.[18]

Hooke attempted the experiment, reporting in a letter to Newton of 6 January 1680 that

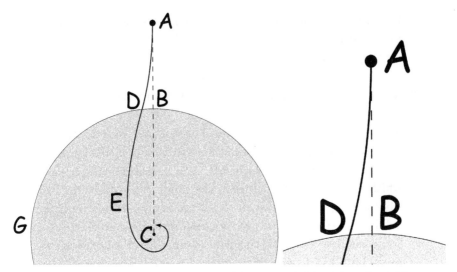

Figure 8.3.   Left: Diagram based on a sketch by Newton in his 28 November 1679 letter to Robert Hooke, showing that an object dropped from a high tower on a rotating Earth should move eastward ahead of the tower as it fell (original available in Lohne 1968, 74). Here the Earth rotates counter-clockwise, and the falling object drops not from A to B directly below, but from A to D, a point to the east of B. Right: Detail from diagram. This part of Newton's sketch could easily serve as an illustration for Riccioli's anti-Copernican argument number 10, in which a heavy ball dropped from high above the Earth should fall straight to the Earth if the Earth is immobile, but should deflect to the east if the Earth has a diurnal rotation.

I have (with as much care as I could) made 3 tryalls of the experiment of the falling body, in every of which the ball fell towards the southeast of the perpendicular, and that very considerably, the least being above a quarter of an inch, but because they were not all the same I know not which was true. What the reason of the variation was I know not, whether the unequall sphericall figure of the iron ball, or the motion of the air, for they were made without doors, or the insensible vibration of the ball suspended by the thread before it was out. But it being a very noble experiment I shall not leave it before I have made a prooffe free from objections, of which I will send you an account.[19]

Hooke then made two trials indoors, reporting to Newton in a letter of 17 January that they were successful and that

I am now persuaded the experiment is very certaine, and that it will prove a demonstration of the diurnall motion of the earth as you have very happily intimated.[20]

The minutes of the 22 January Royal Society meeting report that

Mr. Hooke shewed the ball, that had been let fall from the hight of 27 feet, and fell into a box full of tobacco pipe-clay, sticking in the clay, upon the surface of which were made lines crossing each other: which shewed the true perpendicular point indicated by the ball, when it hung suspended by a thread from the top, and how much the ball had varied from that perpendicular in its descent towards the South and East: and he explained the manner, how the same was performed in all particulars. It was desired, that this experiment might be made before a number of the Society, who might be witnesses of it before the next meeting. The time appointed was the Monday following at three in the afternoon.[21]

There is no further account of this experiment. Since Hooke was Secretary of the Royal Society, if the experiment had been repeated successfully, then he should have recorded that, so presumably it was not.[22] The theoretical eastward deflection for a 27-foot drop is 0.3 millimeters. The rough calculation mentioned previously yields 0.8 millimeters at the equator, so it is surprising that the experiment was even attempted. There is considerable contrast between Hooke's experiment, made with only a few trials and with little effort to control for the effects of the air, and Riccioli's exceedingly thorough falling body experiments.[23]

Yet even had Hooke approached his and Newton's experiment with thoroughness to match Riccioli, he would not have achieved his goal of experimentally proving Earth's motion. The experiment to detect the eastward deflection of a falling body is surprisingly difficult. In 1790, the Italian physicist Giovanni Battista Guglielmini (1763–1817) began an experiment to drop polished lead balls down the inside of Riccioli's Asinelli tower in Bologna and at the Bologna astronomical observatory. He took measures that would have impressed even Riccioli. He floated the balls in mercury before attaching threads to them, so that if their centers of mass and geometric centers were not coincident, they would fall with their centers of mass downward and presumably not rotate. He observed the balls with a microscope to make sure they were released from perfect rest. To ensure that there were no air currents in the tower, he placed candles in various places and waited until the flames were still before making measurements. All these efforts, and the difficulties involved in having to perform the experiments when conditions were perfect, meant he was working on the experiment over a period of more than a year. Nonetheless, he found that the balls, for no apparent reason, would fall rather far apart from one another. He did record an eastward deflection that agreed with theory, but he also recorded nearly as much southward deflection as eastward, which cast doubt on the validity of all the results.[24] The Jesuit astronomer William F. Rigge (1857–1927), in a review of Guglielmini's experiments, wrote that his results could not be considered as even a qualitative detection of the Earth's rotation.[25]

In 1802, Johann F. Benzenberg (1777–1846) also attempted the experiment, in Hamburg, Germany, by dropping balls from a height of 235 feet. Benzenberg recorded an eastward deflection in agreement with theory. He also recorded a smaller southward deflection of 3.4 millimeters, which vanished in a later experiment. Moreover, the extreme values of his measurements varied by as much as nine times their average value, and he detected significant changes in his plumb line caused by the appearance of the sun from behind clouds. Rigge concluded that Benzenberg's results also could not be considered as even a qualitative detection of the Earth's rotation.[26]

Three decades later, in 1831, Ferdinand Reich (1799–1882) took up the experiment, dropping balls down a mine shaft 520 feet deep near Freiberg, Germany. The theoretical eastward deflection was over an inch, and Reich's average experimental deflection was almost exactly that. Although Rigge and others cite Reich's work as an example of a successful detection, his contemporaries noted that his results varied significantly. Still later, experiments in the 1840s in Cornwall with a 1300-foot drop produced wild results, including southward deflections of 10 to 20 inches, while the theoretical eastward deflection is less than 5 inches.[27]

Thus the question of the deflection of falling bodies was still being investigated and discussed at the turn of the twentieth century. During this time Edwin H. Hall (1855–1938) dropped nearly one thousand balls down a tube in his laboratory at Harvard University, seeking to determine exactly how a falling ball behaves, while Florian Cajori (1859–1930) wrote about a falling ball's mysterious southward deflection. Hall, too, noted that "curious things" could occur in these experiments. More prosaic examples of such curious things in Hall's day included magnetic fields deflecting steel or iron plumb lines from the perpendicular, and forces caused by magnetically induced electric currents in conducting metal balls. More extreme examples included plumb lines in a deep mine shaft that were observed not to be parallel, and balls falling down the same mine shaft that were observed never to reach its bottom![28]

Today, when turbulence and chaos are familiar concepts in physics, physicists are not surprised that dropping heavy balls from a tower fails to reveal the Earth's rotation. Nor are they surprised that a ball falling through a great distance deviates unpredictably from its expected path, and that two identically dropped balls do not land in the same place. But even a century ago, the recalcitrant nature of this delicate but apparently straightforward experiment must have appeared to have been "curious" indeed—and all the more so in Riccioli's or Hooke's or Guglielmini's time, when such things as induced magnetic effects were unknown.

This raises an interesting question regarding Riccioli's Coriolis effect argument against the Earth's diurnal rotation. Riccioli supposed that Earth's rotation should reveal itself in an easterly deflection of a heavy falling body. Newton wrote that the "advance of the body from the perpendicular eastward" will be very small, but "enough to determine the matter of fact." This simple test, which appears to involve nothing more than the initial velocity of the ball, the downward pull of gravity, and the upward drag of air resistance, should discover (as Newton wrote) the Earth's diurnal rotation. Yet it is bedeviled by all sorts of "curious things" seemingly beyond explanation. Absent the sort of Herculean effort made by Guglielmini, Benzenberg, Reich, and Hall—the kind made by scientists who know an effect *must* exist and are determined to find it—this simple test will fail to detect Earth's motion.

Moreover, Riccioli's cannon test will also fail to detect Earth's motion even if performed with perfectly accurate cannons. The Coriolis effect turns out to be independent of direction at any given location. A hurled cannon ball will be deflected by the same amount, in the same manner (that is, to the right in the Northern Hemisphere), regardless of the direction toward which the cannon is aimed. Contrary to Riccioli's argument, a cannonball hurled eastward in the Northern Hemisphere will be deflected to the right (southward) by precisely the same amount as its cousin fired northward will be deflected to the right (eastward).[29] It would have been impossible for Riccioli and Grimaldi to determine whether Earth rotates by comparing the deflection of cannonballs fired in different directions. Yet Riccioli and Grimaldi's work predated the Newtonian mechanics and advanced mathematics needed to fully understand the Coriolis effect and how it is independent of direction at any given location. Absent such tools, it is difficult even to visualize the Coriolis deflection of a projectile launched east or west, let alone determine it to be identical with the deflection of a projectile launched north or south. Even today, elementary discussions of the Coriolis effect typically consider, as did Riccioli and Grimaldi's, only projectiles launched north or south.[30] Moreover, while the Coriolis effect is independent of direction *at any given location*, it decreases with decreasing latitude, to zero at the equator. Thus, on the equator, Riccioli is right (at least in theory). This is because a ball launched east or west from the equator remains on the equator, with no effect present, and does not deflect; by contrast, a ball launched from the equator toward a pole leaves the equator, and thus becomes subject to the effect, and deflects (obviously this would only work for a cannon whose range was many hundreds of miles). It seems improbable that anyone would have recognized all this before Newton. Indeed, the mathematics of the Coriolis effect were worked out by the French mathematician and physicist Gaspard-Gustave de Coriolis (1792–1843) in the early nineteenth century. Thus Riccioli's argument that accurate gunners on a rotating Earth should notice a difference between shots fired to the north or south, versus shots fired to the east or west, would seem quite solid in 1651.

Common motion could not defend against these anti-Copernican arguments. If Earth moved, that motion *should* be detectable. But while this line of argument was strong and could not be answered by appeal to common motion, it could be answered by appeal to inadequacy of experiment (much as Galileo had done regarding the cannons fired at targets to the east and west). In other words, Riccioli could be entirely right about the effects he (and Grimaldi; he credits Grimaldi with the physics of the cannon argument)[31] foresaw being present on a rotating Earth, and yet the effects simply might not be detectable by the experiments of the day. No angel was available to help with dropping massive balls from on high, so if the available towers were not tall enough,[32] and if the available guns were not accurate enough, nor of sufficiently long range (not able to hurl their projectiles with "sufficient violence" as Riccioli would have said), then the effects might be "insensible," even if present. Common motion might not defend the Copernicans from this line of argument, but the limits of experimentation might. And while Riccioli believed, for example, that gunners would have detected the deflection of shots to the north, he left room for the possibility that they might not.

But among the 126 arguments there was an anti-Copernican argument even stronger than the "Coriolis effect" argument. This was an argument against which Riccioli believed the Copernicans had no defense. It was Tycho Brahe's "principal argument"—the issue of star sizes, fully updated for the age of telescopic astronomy.

9 The Telescope against Copernicus

Few of the 126 arguments for and against the Copernican hypothesis that Giovanni Battista Riccioli reviewed in the *New Almagest* were decisive. Whether pro-Copernican or anti-Copernican, they could be answered by the other side in the debate. The Coriolis effect arguments seen in the last chapter were very strong. But although they showed that Earth's motion should in principle be detectable, it was always possible that the described effects might avoid detection because, like the supposed effects of annual parallax, they were so small. But among the anti-Copernican arguments were some that were based on a line of argument against which there truly was no satisfactory response—Tycho Brahe's powerful star size objection to the Copernican hypothesis, now reinforced with telescopic observations. This is the line of argument to which Riccioli devotes the most attention—several anti-Copernican arguments (numbers 65 through 70), the response to one pro-Copernican argument (number 9), and at least two chapters (chapter 11 of book 7, chapter 30 of book 9) are all connected to the question of stars, parallax, and star sizes.

Riccioli updated Tycho's argument about star sizes for the age of telescopic astronomy. As we saw in chapter 4, Galileo and Simon Marius observed the telescopic disks of stars. Riccioli also observed these disks, and he measured them. He reports on this with typical thoroughness in chapter 11 of book 7 of the *New Almagest*. He illustrates the method by which he measured telescopic stellar disks, the type and amount of data he collected, and the conclusions he drew. Riccioli begins by informing the reader that determining the apparent diameter of fixed stars depends upon observing planets and comparing planetary and stellar diameters. He and Grimaldi recorded the shape of Jupiter's disk and Saturn's oval form. They obtained accuracy in doing this, he says, by means of repeated and immediate comparisons between what they drew on paper and what they saw through the telescope—apparently they used a sort of "blink comparison," in which one eye observes the paper while the other observes the image in the telescope, or a similar method. They confirmed what their eyes showed them, by observing both together and separately, and by getting the opinion of an unbiased third party. Working during the turn of the years 1649–50 they recorded the appearance of Jupiter and Saturn— Riccioli provides figures of each (fig. 9.1). The diameter of each was measured. Riccioli refers the reader to elsewhere in the *New Almagest* for details on measuring the diameter of a planet. The diameter he reports for Jupiter is 44 seconds of arc (44/3600 of a degree)—a reasonable value, although a bit high compared to what modern calculations give for that date. Riccioli and Grimaldi then divided the diameter of Jupiter (BZ in the figure) into 200 parts; the scale in the figure works out to 100 parts per Roman inch on the Jupiter figure. Using the planetary figures as references, they went on to observe stars, turning to look at the drawings frequently to determine the telescopic size of each star, measured in what is essentially hundredths of a Jovian radius. Riccioli says they got other Jesuits to do this as well, especially Fr. Paulo Casato and Fr. Mattheo Taverna, who had "sharp eyes and minds." They were pleased to find close agreement in their various estimations. Anyone with a good telescope and unbiased observers can use this method to reproduce the results of his observing team, Riccioli notes. The telescopic star diameters the team obtained were recorded in units of hundredths of a Jovian radius, and calculated in terms of seconds and thirds of arc (see fig. 9.2). These sizes range from 18 seconds for Sirius, the brilliant "Dog Star" and the most prominent star in the sky, down to just over 4 seconds for Alcor, a small star in the handle of the Big Dipper (the tail of the Great Bear).[1]

Riccioli next proceeds to comment on the telescopic star sizes observed by others—in particular by Hortensius (Martin van den Hove), who measured the diameter of Sirius to be 10 seconds and the diameters of other

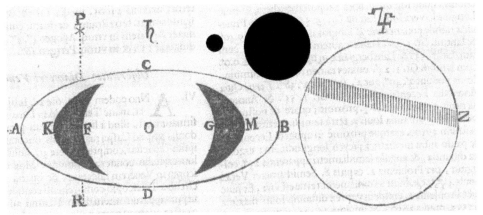

Figure 9.1. Riccioli's Jupiter and Saturn figures (Riccioli 1651, 1:712). Also shown for purpose of comparison are representations of the sizes of Sirius (large black disk) and Alcor (small black disk) as measured by Riccioli and reported in his tables. Image credit: ETH-Bibliothek Zürich, Alte und Seltene Drucke.

I. TABVLA. OBSERVATIONES DIAMETRI Apparentis Fixarum, comparatæ cum Difco Saturni folius quando erat 35″. & Iouis quando 44″. fed illius diametro diuisâ in 160. huius in 200. particulas			
Stellæ Fixæ	Qualiũ Diamet. ♄ 160 ♃ 200	Ergo Apparens Diameter eft *II.* *III.*	Ordo antiquⁱ Magnitudinis
Sirius	82	18 0	1
Lyræ lucida	79	17 24	1
Arcturus	76	16 42	1
Capella	73	16 8	1
Aldebaran	70	15 24	1
Spica	68	15 5	1
Regulus	64	14 5	1
Regel	62	13 40	1
Fomahant	61	13 25	1
Antares	60	13 12	1
Hydra	58	12 45	1
Cauda ♌	57	12 30	1
Procyon	56	12 20	2
Aquila	50	11 0	2
Orion. cingul.	40	8 50	2
Coronæ lucida	38	8 21	2
Polaris	36	7 54	2
Medufæ Caput	32	7 3	3
Propus	28	6 10	4
Pleias lucidior	24	5 16	5
Alcor	20	4 24	6

Figure 9.2. Riccioli's table of the telescopic sizes of stars, from the *New Almagest* (Riccioli 1651, 1:716). The first column to the right of the star name gives its size measured in hundredths of an apparent Jovian radius, the second column in seconds and thirds of arc. The third column is the magnitude of the star. The largest star is Sirius at top; the smallest is Alcor at bottom. Image credit: ETH-Bibliothek Zürich, Alte und Seltene Drucke.

first magnitude stars to be 8 seconds. Riccioli blames the discrepancy between his values and those of Hortensius on two things. The first is a matter of equipment and method. Hortensius had a poor telescope that did not magnify Jupiter much, and the process is sensitive to small errors of measurement and changes in the size of the image of Jupiter (owing to its location relative to Earth at any particular time). The second is a matter of bias. Riccioli thinks that Hortensius, as a Copernican, is biased toward supposing the stars to be smaller. He notes how in the Copernican world system, the stars must be sufficiently distant that there is no perceptible annual parallax caused by the Earth's motion, and how larger star sizes seen through the telescope translate, at Copernican distances, into physically giant stars. The Copernican hypothesis is far less supportable, he says, if the physical sizes of the stars dwarf the orbit of the Earth. However, as of course Peter Crüger might have noted (see chapter 4), and as Riccioli will implicitly note shortly in his criticism of Galileo, the difference between Riccioli's 18 seconds diameter for Sirius and the 10 seconds diameter he ascribes to Hortensius is not sufficient to make a difference in the argument. Also, Riccioli's measurements and those of Hortensius are probably in fair agreement. According to modern calculations, Riccioli probably overestimated the telescopic size of Jupiter, and since he measured stellar diameters in terms of Jupiter, this would inflate his star measurements.

Riccioli also discusses the work of Philips Lansbergen (recall from chapter 5), who cites nontelescopic measurements of star sizes, including those of Tycho, that put the observed diameter of first magnitude stars at a minute (1/60 of a degree, or 60 seconds) or greater, but who then adds that through the telescope star diameters appear much smaller.[2] Riccioli states that the telescope, in "exposing the disks of the stars and scraping off the adventitious ringlets [*cincinnos*] of the rays"[3] is more trustworthy than the naked eye or arbitrary estimation (here we see again the concept of adventitious rays—recall from chapter 4).[4]

Now Riccioli turns his attention to Galileo. He remarks on how the *Dialogue* purports to answer the star size objection to the Copernican system, noting that, even though Galileo says in the *Dialogue* that first magnitude stars do not exceed 5 seconds in diameter as seen through a telescope, Galileo still fails to solve the problem that he is claiming to answer. Riccioli illustrates this with his own calculations of star sizes, showing how even little Alcor (whose telescopic diameter he measures as being smaller than 5 seconds) must be very large if Copernicus is correct. In the *Dialogue* Galileo had also discussed the use of suspended cords to verify his telescopic star diameter measurements, and Riccioli also points out the inadequacy of this method. He is harsh in discussing Galileo: he seems to mock Galileo by echoing language from Galileo's discussion of this issue in the *Dialogue*, and he uses terms like "fallax," "falsitate," and "falsa" both in the text of the paragraph and in the accompanying marginal note. These words convey that Galileo was not merely mistaken, but deceitful. He also mentions Kepler, and how before the invention of the telescope Kepler said stars had large disks while afterward Kepler said the stars were but points.[5]

Riccioli ends chapter 11 of book 7 by advocating following the evidence gathered by means of the telescope. He says he is providing tables (see fig. 9.3) that show the physical sizes of stars calculated from telescopic observations and different estimates of distances to the stars (made by both geocentrists and Copernicans; his calculations are of the same sort as Crüger's), in order that they will be available for any discussion. He says he wants them to be available especially for any discussion regarding the hypothesis of Copernicus.[6]

For in these tables, Tycho Brahe's old star size objection to the Copernican system is on display in full force. Under calculations based on a geocentric hypothesis, the physical size of a small star like Alcor is roughly comparable to that of Earth. By contrast, under calculations based on the Copernican hypothesis, Alcor is enormous—half the diameter of Earth's orbit. Indeed, based on the distance estimates of one Copernican, Godefroy Wendelin (1580–1667), the diameter of little Alcor would measure almost 13,000 times the radius of the Earth. By way of comparison, Tycho had calculated the distance to the stars to be 14,000 Earth radii. Thus this one single little star would conceivably be a significant fraction of the size of the *entire Tychonic universe*. The much larger Sirius would easily be larger than Tycho's universe.[7]

Why is this true? After all, the telescope, in getting rid of the adventitious rays, revealed the stars to be much smaller than they appeared to the naked eye (see chapter 4). Would this not result in stars being physically smaller?

II. TAB. MAGNITVDO VERA FIXARVM MAXIMÆ ideft SIRII, & MINIMÆ, cuiufmodi eft ferè Alcor: Pofita Diametro obferuata à nobis in Sirio 18″. in Alcor 4″. 24‴. & Diftantia Fixarum afferta in Hypothefi TERRÆ QVIESCENTIS.					
Auctores	Diftantia Fixarum à Terra	Magnitudo Vera Sirij feu Canis Maioris		Magnitudo Vera Alcor, quæ eft prope mediam caudæ Vrfæ maioris	
Diftantiæ	In femidiametris Terræ rotundè	Diameter vera habet Terræ Diam.	Corpus côtinet Terram vicib.	Diameter habet Diamet. Terræ	Corpus continet Terram vicib.
Tycho Ptolemaici Nos	14000 Max. 40000 210000	0 3 17	61/100 ¼ ½	0 42 5355	¼ 0 0	0 0 4	1 1/100 86/100 0	0 0 64	3/10000 1/... 0

III. TAB. MAGNITVDO VERA FIXARVM MAXIMÆ & MINIMÆ Ideft Sirij & Alcor: Pofita diametro obferuatà à nobis in Sirio 18″. & in Alcor 4″. 24‴. ac Diftantia Fixarum afferta à Sectatoribus HYPOTHESIS COPERNICANÆ					
Auctores	Diftantia Fixarum à Terra in	Magnitudo Vera Sirij feu Canis Maioris		Magnitudo Vera Alcor in cauda Vrfæ Maioris	
Diftantiæ	Semidiametris Terræ	Diameter habet Terrę Diametros	Corpus continet Terram vicibus	Diameter habet Diametros Terræ	Corpus continet Terram vicibus
Hortenfius	10,312,227	899	726,572,600	442	86,355,888
Galilæus	13,046,400	1138	1,473,760,072	558	173,741,112
Lansbergius	41,958,000	3658	48,947,466,312	1796	5,793,206,356
Keplerus	60,000,000	5232	143,219,847,228	2568	16,933,994,432

Figure 9.3. Riccioli's tables from the *New Almagest* (Riccioli 1651, 1:716–17) show-ing the calculated physical sizes of Sirius (center two columns, giving size compared to Earth by diameter and by volume) and Alcor (right two columns) under various scenarios, based on his telescopic measurements of their apparent diameters (18 sec-onds for Sirius, 4 and 24/60 seconds for Alcor; see fig. 9.2). When calculated under the scenario of a fixed Earth (Table II, top), Sirius ranges from a fraction of the diameter of Earth up to about 17 times the diameter of Earth, depending on whose distance es-timates are used for the stars. When calculated using stellar distance estimates from various Copernicans (Table III, bottom), even Alcor becomes quite large. When calculated using the distance necessary for the stars to have a parallax of less than 10 seconds of arc (Table IV, next page), little Alcor balloons to over 20,000 times Earth's diameter under one estimate (based on distances calculated from the work of *Vendelinus*, that is, Godefroy Wendelin), and over a trillion times Earth's volume. By contrast, the distance to the stars themselves is only 14,000 times the radius of Earth according to Tycho (Table II, top left). Thus, at least if the estimates of Wendelin are correct, one very minor star is, under the Copernican hypothesis, conceivably compa-rable to the size of the entire Tychonic universe. It should be noted that Wendelin had calculated the Sun itself to be more distant, and therefore larger, than was commonly accepted, and so if the Sun were the standard of comparison the numbers would be less dramatic. Nonetheless, the calculations show Tycho's star size problem to remain very much in force under telescopic observations.

These tables are plagued by typographical errors. Most noticeable of these is the *Galilaeus* entry for the volume of Sirius in Table IV, which is missing a zero and thus too small by a factor of ten. However, the difference between the geocentric and Co-pernican star sizes are so great that even errors do not really change the impact of the table. See Graney 2010b, 460–61, for a translated and corrected version of these tables.

Images credit: ETH-Bibliothek Zürich, Alte und Seltene Drucke.

IV. TAB.   MAGNITVDO VERA FIXARVM MAXIMÆ & MINIMÆ
Ideft Sirij & Alcor: Pofita diametro apparenti obferuatâ à nobis in Sirio 18″. in Alcor 4″. 24″. & Diftantia
afferendâ à Copernicanis, fi velint Parallaxim Fixarum factam à motu Terræ annuo non
excedere 10″. & tueri Diametrum Orbis Annui ab ipfis pofitam .

Auctores	Diftantia Fixarum Afferenda	Magnitudo Vera Sirij feu Canis Maioris		Magnitudo Vera Alcor in cauda Vrfæ maioris	
Diftantiæ	Semidiametri Terræ	Diam. habet Terrę Diam.	Corpus continet Terram vicibus	Diam. habet Terræ Diam.	Corpus continet Terram vicibus
Copernicus	47,439,800	4170	71,677,713,000	1992	4,378,454,048
Herigonius	49,502,400	4350	82,3 12,875,000	2068	8,844,058,432
Galilæus	49,832,416	4380	8,427,672,000	2092	9,155,562,688
Bullialdus	60,227,920	5300	148,877,000,000	2530	15,941,277,000
Lansbergius	61,616,122	5424	159,371,956,024	2588	17,333,761,472
Keplerus	142,746,428	12550	1,976,656,375,000	6000	216,000,000,000
Vendelinus	604,589,312	53200	15,056,882,800,000	25380	1,767,384,872,000

Fundamenta harum diftantiarum vide in lib. 6. cap. 7. num. 15.

The answer to these questions is annual parallax. Tycho measured prominent stars to be 2 minutes of arc, or 120 seconds, in diameter; the telescope reduced that by a factor of roughly ten, to 12 seconds, more or less. But recall that Tycho also could measure the position of a star to within a minute of arc, or *60 seconds*. If a star had an annual parallax exceeding 60 seconds, he could detect it, and based on the fact that no star showed a detectable parallax, he could calculate a minimum possible distance to the stars under the Copernican hypothesis. Riccioli's tables are calculated based on annual parallax not exceeding *10 seconds*. The telescope allows for a more sensitive measurement of parallax: 10 seconds rather than 60. This increases that minimum possible distance to the stars in the Copernican system. Thus while the telescope shows stars to be smaller in apparent size than what nontelescopic instruments show, it also forces them to be farther away. Smaller apparent size means smaller stars; farther away means larger stars (see fig. 9.4). In Riccioli's tables, the two effects are seen to largely cancel out. The result is that Tycho's star-size objection stands firm.

Interestingly, the Copernican Hortensius (who translated Lansbergen's *Considerations*; see chapter 5), to whose telescopic star diameter measurements Riccioli makes reference, largely agrees with Riccioli. Hortensius's measurements and discussion are found in his 1633 *Dissertation Concerning Mercury Seen on the Sun*, where he notes that with considerable effort of eye and mind he determined Sirius to have a telescopic diameter of 1/6 of a minute of arc and, noting that not all stars of the same magnitude have the exact same diameter, he lays out a table of general diameters for

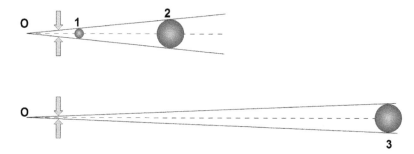

Figure 9.4.   Illustration of the effect, or lack thereof, of the telescope on the star size question. Recall from figure 3.8 that the farther distant a star may be from an observer at O, the larger its physical size must be (2 vs. 1 in the upper figure). Compared to a nontelescopic instrument, a telescope yields a smaller apparent size for a star, as indicated by the arrows in the lower figure. Were the star at the same distances as 1 or 2 in the upper figure, the telescope would indeed reveal it to be smaller in physical size. However, the telescope, because it increases sensitivity to parallax, requires that the star be yet more distant (at 3). The result is that the star, even though it may have a small apparent size, still has a very large physical size.

each magnitude of stars. He then uses these diameters, and a maximum parallax of just under 15 seconds (a little greater than Riccioli's 10 seconds; he references the value to Lansbergen), to calculate the physical sizes of stars. As Hortensius's measurements of star diameters are a little smaller than Riccioli's, and his minimum parallax considerably larger, the results of his calculations are that no star is physically larger than the orbit of Earth. Nonetheless, the stars are by no means small. Certainly they are far larger in size than the Sun (recall how Crüger had shown that if Copernicus was right, even a star with diameter as tiny as a single second would dwarf the Sun). Hortensius acknowledges the problem of the vast physical sizes of stars and how these sizes may lead people to reject the motion of the Earth.[8] The *New Almagest* was published nearly twenty years after Hortensius's *Dissertation,* so it might be more correct to say that Riccioli largely agrees with Hortensius rather than the other way around: both the Copernican Hortensius and the Tychonian Riccioli agree that the telescope shows stars to be huge under the Copernican hypothesis.[9]

Whereas Riccioli grants that the Coriolis effect arguments could be answered by claiming that experiments simply had not detected the various effects involved, he makes no such concessions regarding the star size arguments. The only answer the Copernicans have to the star size argument, he says, is their appeal to the power of God.[10] Of course, who can deny the power of God? The Copernican response was, in that sense, irrefutable. Still, Riccioli rejects invoking the power of God as a response to the star size argument, stating that "even if this falsehood cannot be refuted, nevertheless it cannot satisfy more prudent men."[11]

The Copernican use of a theological answer to respond to a scientific argument is unacceptable to Riccioli.[12] He devotes chapter 30 of book 9 to discussing the star size question, and the Copernican response to it, in some detail. Copernicans can deny this is a problem, he says, via claiming the vastness of the stars as evidence of Divine Magnificence: God made the Sun far larger than the Earth, so likewise the stars are larger than the Sun. A Copernican can also compare the vast sizes the stars must have in the Copernican hypothesis to the speed the stars must have in whirling about the Earth if the Earth is fixed, and then declare the Copernican side to be much more credible. But none of this reasoning, says Riccioli, is satisfactory.[13]

If God's intention was for the vastness of the stars to show His Greatness to mortals, Riccioli writes, while setting them so far distant from us that their sizes appear barely one hundredth the apparent diameter of the Sun, well,

> Surely He might have provided a few other pieces of information that would have allowed us to come into certain knowledge of this distance and size. Yet he did not, as in fact all Astronomical phenomena can be saved [i.e., fully explained] without the Copernican hypothesis. Moreover, experiments involving the physics of gravity and projectiles clearly contradict that bogus hypothesis.[14]

There is no scriptural support for the Copernicans' theological response to the star size question, according to Riccioli. He notes that God Himself, speaking through the Psalmist, describes the Sun as moving:

> He hath set his tabernacle in the sun: and he, as a bridegroom coming out of his bride chamber, Hath rejoiced as a giant to run the way: His going out is from the end of heaven, And his circuit even to the end thereof: and there is no one that can hide himself from his heat.[15]

Nowhere does scripture mention stars having sizes that vastly exceed the
Sun and the Earth, says Riccioli—it only mentions their innumerable
multitude. It describes only one Sun—the giant in the aforementioned
Psalm. He notes that scripture also says,

> The firmament on high is his beauty, the beauty of heaven with its
> glorious shew. The sun when he appeareth shewing forth at his ris-
> ing, an admirable instrument, the work of the most High. At noon he
> burneth the earth, and who can abide his burning heat?[16]

Apparently, says Riccioli, to the Copernicans this "giant" is a pygmy, this
"admirable instrument" is a contemptible little tool, by comparison with
the fixed stars.[17]

As for the argument that the vast sizes of the fixed stars in the Copernican hypothesis is no worse a problem than the great speed of the
fixed stars in geocentric hypotheses, Riccioli essentially notes that either
the Earth rotates or the stars rotate diurnally—and in either case the
motion is proportionately the same: one circumference per day. Only because we measure the circumferences in our units do we say that there is
a difference between the two.[18] Riccioli also points out that the Copernicans should not have a problem with awesome motion of the fixed stars,
seeing as how they are so inclined toward invoking Divine Omnipotence.
After all, if God did want to point out His Magnificence to mankind by
way of the heavens, would not this awesome motion be a more apparent
manner of doing so than by hiding that Magnificence in apparently tiny
stars whose awesome size we can only deduce by means of the Copernican hypothesis?[19]

As in the time of Tycho Brahe and Christoph Rothmann, the problem of the sizes of stars implied by the Copernican hypothesis, and the Copernican tendency to appeal to God's power to answer that problem, were
the most decisive arguments against that hypothesis. Riccioli had updated
Tycho's argument for the age of the telescope, making measurements of
stars through a procedure that others could use to replicate those measurements. All astronomers with good telescopes could follow Riccioli's
procedure and see the problem for themselves. The deflections of cannon
balls might be too small to measure, but the stellar diameters, and the lack

of observable annual parallax, could be measured by anyone with a good telescope who followed some basic procedures. Any astronomer who did so could then either reject Copernicanism and become Tychonic, or accept Copernicanism and believe that the Sun and its tiny planetary system were surrounded by a vast universe of giant stars, and maybe justify those huge bodies by saying that they spoke of God's power, or that they were perhaps His mighty warriors. That was, of course, possible; one could never refute the power of God. But for the more prudent, the star size argument was a decisive blow against the Copernican system.

10   It Can No Longer Be Called
        "False and Absurd"

History has not been kind to the anti-Copernicans. Consider that well-known sort of "modern morality play in which brave Reason suffers at the hands of villainous Superstition before triumphing in the sunny dawn of Science."[1] In our play, the anti-Copernicans are clearly on the side of villainous Superstition, while the Copernicans are on the side of brave Reason. After all, since Galileo's *Dialogue* "masterfully demonstrates the truth of the Copernican system . . . proving, for the first time, that the earth revolves around the sun"[2] (or so we are told), to have been anti-Copernican was to have been anti-reason and anti-science. We have heard this all before: those in authority "believed that faith should dominate; and Galileo believed that truth should persuade,"[3] and so forth. As for the anti-Copernicans, well, they relied on counting up arguments, on philosophical and scriptural authority, on decrees from Rome, and so forth: "Riccioli had no real arguments to support the geocentric system other than the Bible and the authority of the Church."[4]

This play's narrative is continually challenged by scholars of the history of astronomy, who have long been well aware that it is false insofar as the scientific evidence in the sixteenth and seventeenth centuries was by no means strong enough at the time to demonstrate the truth of Copernican heliocentrism over hybrid geocentrism. For example, Kerry Magruder has recently written that given the diversity of cosmological views circulating in the mid-seventeenth century, it seems . . . misleading simply to characterize that debate as the collision between the Copernican and the Aristotelian/Ptolemaic worldviews—although this rhetorical trope was famously employed by Galileo in his 1632 *Dialogue on the Two Chief World Systems*. When Galileo wrote that dialogue, the Ptolemaic system already had been set aside, at least among mathematical astronomers. . . . Copernicanism was admired as the standard by which the mathematical aspects of other systems were judged, but alternatives proliferated rapidly as the search for observable distinguishing evidence bogged down. Transformations of systems threw all in doubt. Some systems were not just empirically similar, but geometrically equivalent. . . . Observations to empirically distinguish between a multiplicity of systems proved elusive. . . . What we now disparagingly refer to as "hybrid systems" were not regarded as short-term compromises with an inexorably-advancing Copernicanism, but as provisional experiments that seemed at least as warranted as the Copernican extreme. The Scientific Revolution is far more interesting than a conflict between two chief world systems.[5]

But let us grant that reducing the narrative to a conflict between two chief world systems—Aristotelian/Ptolemaic geocentrism, backed by authority, tradition, and religion, versus Copernican heliocentrism, backed by the telescope, reason, and science—is painting with too broad of a brush, and neglects many details historians have revealed. Even so, is not the morality play's narrative ultimately correct? Religion called heliocentrism "foolish and absurd," and yet today we know the Earth to circle the Sun. Is not the bottom line this: science backed heliocentrism, and religion backed geocentrism?

In light of Tycho and Riccioli, Rothmann and Lansbergen, the answer to that question is clearly, "No." With the publication of the *New Almagest*, an array of arguments from physics and astronomy supported hybrid geocentrism. These arguments apparently had no answer at the time.

It is difficult to imagine a situation in which they could be answered, granted the knowledge then available. To illustrate this, let us imagine that the year is 1655, and . . .

In the wake of the publication of the New Almagest, *a coordinated international scientific effort is underway to determine, once and for all, whether or not the Earth orbits the Sun! Precision experiments are being conducted on falling bodies! The most skilled gunners and the best gun-makers are volunteering in the scientific effort! They are testing the accuracy of weapons when fired at targets to the north, east, south, and west. The finest telescopes are being constructed! They are being honed and tuned to remove the adventitious rays from even the brightest stars. Apparent stellar sizes are being measured with unprecedented precision! An international observing campaign is being launched for the purpose of seeking out double stars. These will be used to probe for annual parallax. Other manifestations of annual parallax are being sought as well.*

Years pass. . . .

The results that are coming in are not looking good for the Copernicans. The falling bodies experiments have been inconclusive. The experimenters cannot get balls to drop to a consistent spot. There is no clear reason for this. The wide variety of landing points makes the measurement of any average deviations from the path of a plumb line completely untrustworthy! Profit-seeking, sensationalist elements in the popular press are loudly speculating that mysterious diabolical influences may be disrupting this promising experiment. There are dark rumors that scientists are dabbling in things men were never meant to know!

The cannon experiments are likewise inconclusive. No differences are being found between shots fired north or south, and identical shots fired east or west. What these experiments are demonstrating, however, is the inaccuracy of artillery. The notion that gunners can place a shot right into the mouth of an enemy cannon has been shown to be pure myth! It seems perhaps a few lucky shots remained too long in memory and became accepted lore. A scandal has erupted! According to sources who wish not to be named, one overly boastful general, who bragged that his men would quickly solve a problem that certain Jesuits could not, has been forced to resign his position. But the inaccuracy issue has negated the argument that gunners, on account of their great skill, would have already noticed a deflection effect if it existed. We know now that the gunners do not have such great skill!

The campaign of telescopic observations seems to be producing re-
sults. Quite a few double stars are being found. Not one has revealed any
annual parallax.[6] *Efforts to detect other manifestations of annual paral-*
lax have also been unsuccessful. The failure of this careful search for par-
allax means that, if Copernicus is correct, the distance to the stars must
be even more vast than previously estimated! Skilled measurements of
stellar diameters have yielded values smaller than what Riccioli meas-
ured, but larger than what Galileo obtained.[7] *Combined with the new,*
vaster Copernican distances to the stars, these apparent sizes again are
translating into stupendous physical sizes comparable to or greater than
that of the orbit of Earth in the Copernican hypothesis!

More years pass. . . .

The international effort is ending in failure and discord. Despite the
research, the situation remains largely unchanged from what it was
when Tycho and Rothmann debated at the end of the sixteenth century!
The motion of the Earth ought to be detectable, assertions to the con-
trary regarding common motion aside. Experiments with projectiles and
falling bodies can, in principle, be designed that should detect, or not de-
tect, the motion of the Earth. But in fact all such experiments have been
inconclusive. Observations of stars have indicated that, if Copernicus is
right, the stars must be amazingly far away, and stupendously large. Ty-
chonic geocentrists are continuing to cite the absurdity of this. Coperni-
cans are again attributing their titanic stars to the Power of God, and
pointing to the harmony and beauty of the Copernican system.

Governments and others who funded the international effort, on
promises that they would be helping to answer the greatest scientific
question of the age, are withdrawing support for the project. Over two
dozen lawsuits related to the project have been filed in courts in several
countries. And public interest in science, which had risen dramatically
as the effort was building, has declined to record lows! Will the question
of Earth's motion ever be answered? That remains to be seen.[8]

Our imagined scenario illustrates that the answer is "No" regarding
the question about science backing heliocentrism and religion backing
geocentrism, because in the middle of the seventeenth century, Tycho's old
anti-Copernican arguments simply could not be answered, except by an

appeal to the power of God. While the Jesuit Riccioli may not have been willing to utterly dismiss God's power, few scientists today would stop at his assessment that such an appeal would not satisfy the prudent. Today a theory that invokes God's power to explain its creation of a whole new class of titanic bodies would be completely unacceptable as a scientific theory. To be fair to the Copernicans, a theory that creates a class of enormous bodies would perhaps not be so problematic in itself. After all, astronomers have done something much like this in theorizing the existence of dark energy and dark matter, which together account for the overwhelming majority of the universe. But dark matter is not explained as being God's palace, or his mighty warriors. Such explanations today would be considered as Tycho and Riccioli considered them: absurd. As we saw in chapter 1, Riccioli asked which hypothesis was true, according to the best arguments: the one that supposes the motion of the Earth, or the one that supposes the immobility of the Earth? His answer was the latter, and it seems likely that science today would agree with his thinking.

Thus it is fair to say that, contrary to our question from a few paragraphs back, science backed geocentrism and religion backed heliocentrism. Or, to put it more correctly, just as science supported the Copernican hypothesis over the Ptolemaic with discoveries such as the phases of Venus, so science supported the Tychonic hypothesis over the Copernican with discoveries such as adventitious rays and the telescopic disks of stars. Likewise, just as religion was used to support geocentrism, by citing scriptural backing for a fixed Earth, so also religion was used to support heliocentrism, by providing a supernatural explanation (also supportable by scripture) for a dramatic absurdity in the Copernican system. The story of the "Copernican Revolution" does not look so much like a morality play about brave reason and villainous superstition, about "science vs. religion," as it looks like a battle between two scientific theories, about "science vs. science," with a little "religion vs. religion" thrown in as well.

It is remarkable that the story of this science vs. science battle has been so thoroughly forgotten. Generally we want the historical narrative to agree with the historical record. Yet the reader who peruses secondary sources for information on Riccioli and his arguments will find (if he or she finds much at all) comments to the effect of those listed in chapter 7: the supposed measuring of evidence "not by the weight but by the number of the arguments," the "egregious structure" of Riccioli's analysis, his lack of arguments to support the geocentric system "other than the Bible and the authority of the Church," and so on. Moreover, the only author to attempt even a partial listing of the 126 arguments, the nineteenth-century historian of science Jean Baptiste Joseph Delambre, does not distinguish between those arguments to which Riccioli believes there is a valid answer, and those to which Riccioli believes there is no valid answer. Delambre grants but a single line to the all-important argument concerning the size of the stars—"Diameters, motions and distances of the Fixed stars: Nothing is certain on either side."[9] However, that is one line more than the reader will find on the subject from most any other secondary source.

One might think that Catholic sources would do better in this regard. After all, the role of villainous superstition in the aforementioned morality play is generally assigned to the Roman Catholic Church. And even though an old Catholic publication cites the *New Almagest* as being "the most important literary work of the Jesuits during the seventeenth century,"[10] Catholic and Jesuit sources have forgotten Riccioli and his titanic stars no less: the *New Catholic Encyclopedia* has no entry for Riccioli.[11] Indeed, David Wootton has recently argued that Pope Urban VIII gave Galileo permission to reopen the debate on heliocentrism and was willing to see a new science triumph.[12] Might the *New Almagest* (or at least book 9 of it) be the book Pope Urban VIII was hoping Galileo would write?[13] Such speculations aside, the story of Tycho and Riccioli, upholding a scientific approach to the universe against heliocentrists (such as Digges, Rothmann, Lansbergen, and even Copernicus himself) who viewed as the glorious work of God the vastness of their universe and the titanic stars within it, should not be a forgotten part of Catholic history. It is a story that is relevant to modern discussions regarding science and religion. It illustrates how different are modern assumptions from what was thought in the time of Tycho and Riccioli, when geocentrists employed science and the telescope against an "absurd" and "fraudulent" theory that required Divine Omnipotence to solve its scientific problems.[14]

Catholics are hardly the only ones who might have a particular interest in this story—we might suppose scientists would also have been disinclined to forget Riccioli and Tycho. Riccioli and Tycho illustrate how difficult it can be to get answers in science, and how science is a complex and nuanced undertaking where "the answers" can elude even good scientists. This important idea is difficult for scientists to convey to students and to members of the general public, in whom the failure to understand how elusive the answers can be in science sometimes generates some weird ideas. For example, people may think that because science has difficulty determining if things we eat are harmful to our health (caffeine, saccharine), science should simply be dismissed regarding such matters because "in a few years they'll be telling us something else." Far more extreme ideas regarding science are not uncommon, especially the idea that a simple answer is known, but science is hiding it. Here "it" might be the cause of increasing autism rates (vaccines) or the reason we have not returned to the Moon (aliens—unless it is that we never went to the Moon and NASA faked the whole thing in a studio) or why we don't have solar-powered cars (the technology is being suppressed by the energy companies). The vaccine issue, and the problem of public distrust of science in general, has been a recurring topic on NPR's *Science Friday* talk show in the 2010s.[15] In all of these examples there is an implicit rejection of any sense that answers might be challenging and elusive in science. Rather, the assumption is that the answer is known but kept hidden. It would be easier for scientists to dispel the notion that dark and powerful forces conspire to hide scientific truths were the standard story of one of the foundational events of modern science not a modern morality play with just such a conspiracy at its heart. The story of Tycho and Riccioli and the battle of "science vs. science" in the Copernican Revolution shows that answers in science, even answers to what we today think of as the most basic questions, can be highly, and intrinsically, elusive. Moreover, this story also shows that science does make progress, as science did eventually figure out that Copernicus was right. Scientists might have done well to have remembered this story, yet they have remembered it no more than have Catholics.[16]

At this point the reader may well be wondering how science did make progress. If the case for hybrid geocentrism was so strong in the 1650s, how did science eventually figure out that Copernicus was right? As we have seen in chapter 9, the Coriolis arguments were eventually shown to be valid evidence for heliocentrism, starting with the heroic but questionably successful experiments of Guglielmini and continuing through Hall's experiments at the turn of the twentieth century. However, such experiments were always too delicate, and too full of "curious" effects, to produce a clear, powerful demonstration of Earth's motion.

A different manifestation of the Coriolis effect would do that. Consider Riccioli's argument about a cannon ball hurled to the north. The Earth is rotating from west to east. The ball travels from faster-moving southern ground toward slower-moving northern ground, and so the greater eastward velocity it possesses (on account of having been launched from faster-moving ground) causes it to outrun the ground it passes over while going northward: it deflects to the east, that is, to the right. On the other hand, if the ball is hurled toward the south, the ground outruns it, and it deflects to the west—again to the right. Now instead of imagining this occurring in a single large movement to the north or south, imagine it occurring in many small repeated movements to the north and south, such as would occur in the case of a pendulum swinging in a north-south plane. Every northward swing yields a tiny rightward deflection. Every southward swing yields another tiny deflection, again rightward. Each deflection is individually negligible, but as the pendulum swings back and forth the effect of these negligible deflections accumulates, and the plane of the pendulum's swing rotates rightward—clockwise, as seen from above. As we saw in chapter 9, the Coriolis effect at a given location is actually independent of direction, and thus the pendulum is continually directed to the right, not just when swinging in the north-south plane. The result of this is the Foucault pendulum, the standard demonstration of the rotation of the Earth found in museums and college science buildings everywhere (fig. 10.1). Léon Foucault first demonstrated such a pendulum in Paris in 1851. Thus two centuries after Riccioli's 1651 New Almagest discussion of the foreseen effect of the Earth's supposed rotation, a simple, clear demonstration of that effect was finally developed. The Foucault pendulum is also a simple, clear demonstration of the inadequacy of the Copernican common motion argument that said that Earth's motion could not be detected by experiment. The Foucault pendulum demonstrates Galileo's ship cabin analogy to be invalid.

While the Coriolis arguments were valid, and the various effects detected, the all-important star size argument was based on a misunderstanding of the apparent sizes of stars. The disks that seventeenth-century telescopes revealed when turned to the stars were spurious (in contrast to

Figure 10.1. The Foucault pendulum at the Kentucky Science Center in Louisville, Kentucky. Image courtesy of the author.

the nonspurious disks the telescope revealed when turned to the planets—a most confusing situation, as was discussed in chapter 4). As we saw in chapter 5, Francesco Ingoli had suggested to Galileo a possible answer to the star size argument: that the stars were somehow different. The first documented report recognizing evidence for this—evidence for the spurious nature of the apparent sizes of stars (those measured when seen through a telescope)—came from the English Astronomer Jeremiah Horrocks (1618–1641). Horrocks, in a report on the 1639 transit of Venus across the Sun, noted that he and his friend William Crabtree (1610–1644) had once observed the Moon passing through the stars of the Pleiades. The stars had simply winked out as the Moon passed in front of them:

> If with a telescope you observe [the Moon] obscuring the stars with her dark edge, you will see the stars as soon as they touch its border, immediately, and as if in the blink of an eye, vanish. Which W. Crabtree and I have observed most clearly in the appulse of the Moon to the Pleiades on the evening of March 19, 1637.[17]

Horrocks goes on to say that this is evidence that stars have no true size, but are in fact mere points of light,

> for the more perfectly [the telescope] shows the fixed stars, the more it shows them as mere points, which is also evident in the appulse of the Moon to the Pleiades; for the moment the Moon covered the true body of the fixed stars, the false rays vanished instantaneously. If they had been emitted from the true body, they would have vanished by degrees and not at all in a single moment.[18]

Horrocks's comments are intriguing, and a clear record of an astronomer discovering that the telescope was producing spurious information as regards the stars. His report is an accurate reflection of how stars are seen to behave when the Moon passes in front of them. However, he gives this report while seeking to justify the small size of Venus that he recorded during its transit across the Sun. He goes on to discuss planetary sizes as though the planets increase in size in proportion to their distance from the Sun, so that if they were observed from the Sun, they would all have the same apparent size. This seems clearly counter to telescopic observations (especially considering he cites Galileo's star size measurements). Thus his remarks on the sizes of stars and planets are a mixture both of excellent observations and of nonsense obvious to any careful observer.[19]

But more importantly, while Horrocks made his observations of the Moon passing through the Pleiades in 1637, his work was not published until 1662, well after the *New Almagest*. By this time Christiaan Huygens's 1659 *The System of Saturn* had appeared, in which Huygens declared that observations of stars filtered through a smoked glass showed that the sizes of their disks were changed by greater filtering, something that did not happen with the Moon or planets. Huygens would later write,

> Before the Invention of Telescopes, it seemed to contradict *Copernicus's* Opinion, to make the Sun one of the fix'd Stars. For the Stars of the first Magnitude being esteem'd to be about three Minutes Diameter; and *Copernicus* (observing that tho' the Earth changed its Place, they always kept the same distance from us) having ventured to say that the *Magnus Orbis* [the sphere of Earth's orbit] was but a Point in respect of the Sphere in which they were placed, it was a plain Consequence that every one of them that appeared any thing bright, must be larger than the Path or Orbit of the Earth: which is very absurd. This is the principal Argument that *Tycho Brahe* set up against *Copernicus*. But when the Telescopes took away those Rays of the Stars which appear when we look upon them with our naked Eye, (which they do best when the Eyeglass is black'd with Smoke) they seemed just like little shining Points, and then that Difficulty vanished, and the Stars may yet be so many Suns.[20]

Thus shortly after the *New Almagest*, evidence began to accumulate indicating that stars were actually points. Interestingly, a decade and a half after the *New Almagest*, Riccioli in his 1665 *Reformed Astronomy* devotes less emphasis to the issue of the physical sizes of stars. He again publishes his table of telescopically measured star diameters, but without the extensive tables of and commentary on the physical sizes of stars as they relate to the Copernican system (although he does mention the issue). He includes a table of telescopic star diameters measured by other astronomers, and quotes from Huygens's *System of Saturn* regarding measuring the diameter of planets. Perhaps after reading Huygens he was less willing to invest time in the star size issue.[21]

Not all astronomers accepted the idea that stars might be points. John Flamsteed (1646–1719), the first English Astronomer Royal, criticizes Huygens's "stars are points" idea in some detail in a 1702 letter. Flamsteed points instead to Riccioli's measurement of Sirius as being 18 seconds in diameter, "a very sensible quantity," he notes. He then writes,

> I suppose I may be allowed to quote an observation of my own, made at Derby, October 22, 1672, in the morning, which I showed Mr. Keile, and which I will give you in the very words I wrote it down as soon as taken.
>
> "1672, October 22. When Mercury was about 10 deg. high, I observed him in the garden with my longer tube (of 14 foot); but could not with it see the fixa [fixed star] (near him), the daylight being too strong; only I noted his diameter 45 parts = 16", or a little less; for, turning the tube to Sirius, I found his diameter 42 parts = 15" which I judged equal to Mercury's. The aperture on the object-glass was ¾ of an inch: so that Sirius was well deprived of spurious rays, and shined not turbulently, but as sedate as Mercury; the limbs of both well defined, but Sirius best."
>
> I was with Mr. Newton on Friday last, and told him of this observation. He would have said something in defence of his friend, from the nature of the difform rays of light; but when I urged the smallness of the aperture on the object-glass, he let his discourse fall. There is nothing to be said against this observation; for the fault of the glass was as great in Mercury as Sirius. If Mercury had a sensible diameter, so had Sirius. If he will say Sirius had none, Mercury must have none; for both were observed with the same glass, and the same aperture on it. Mr. Huygens says this may proceed "*ab aliqua visus fallacia* [from some deceit of seeing]" . . . ; but 'tis plain prejudice, for his own method had prepossessed him: and having showed, then, the reason of his mistake, you ought to pardon him for the sake of his many useful inventions, and excellent treatise *De Horologio Oscillatorio*.[22]

Flamsteed may have been sure that the telescopic disks of stars were as real as those of the planets, but at least one astronomer was not of that opinion. In 1720 the English astronomer Edmond Halley (1656–1742, of "Halley's comet" fame) would write:

> In the Memoires of the *Royal Academie* of *Paris*, for the Year 1717 but now very lately published, there is one very remarkable Essay, by Mr. *Cassini*, concerning the *Annual Parallax* of the *Fix'd Stars*, and particularly of *Sirius*, and in conclusion, he determines the Diameter of *Sirius* to be as much bigger than that of the *Sun*, as the *Sun's* is greater than that of the *Earth*, which he supposes to be 100 times: And the distance from the *Sun* to the *Earth* being certainly about 100 Diameters of the *Sun*, it will follow, that the Globe of *Sirius* must be a Sphere, whose Diameter must equal the distance between the *Earth* and *Sun*.[23]

Halley then relays how Cassini used a reduced aperture on his telescope, "to take off the Spurious Rays of the *Star*, which then appeared round, and sufficiently well defined," and then measured the apparent size of Sirius (5 to 6 seconds is Cassini's value) and calculated its physical size. But after relaying all of Cassini's work Halley adds,

> But before this obtain a full assent, it may not perhaps be amiss to enquire whether the suppos'd visible Diameter of *Sirius* were not an Optick Fallacy, occasioned by the great contraction of the *Aperture* of the *Object Glass*: For we all know that the Diameters of *Aldebaran* and *Spica Virginis* are so small, that when they happen to immerge on the dark Limb of the *Moon*, they are so far from loosing their Light gradually, as they must do were they of any sensible magnitude, that they vanish at once with their utmost Lustre; and emerge likewise in a Moment, not small at first, but at once appear with their full Light, even tho' the Emersion happen very near the *Cusp*; where, if they were four Seconds in Diameter, they would be many Seconds of Time in getting entirely separated from the Limb. But the contrary appears to all those, that have observed the Occultations of those bright Stars. And tho' *Sirius* be bigger than either of them, yet he is by far less than two of them; and consequently his Diameter to theirs is less than the

Square Root of 2 to 1, or than 14 to 10; whence, in Mr. *Cassini's* excellent 36 Foot Glass, those Stars ought to be about four Seconds in Diameter; and they would undoubtedly appear so, if view'd after the same manner; whereas we are *aliunde* [i.e., for other reasons] certain, that they are less than one single Second in Diameter. The great strength of their native Light, forming the resemblance of a Body, when it is nothing else but the spissitude [denseness] of their Rays.[24]

Thus here we see Halley arguing in 1720 that the apparent sizes of stars are spurious. But here we also see, even in 1720, some astronomers still accepting those apparent sizes as real, and from them calculating vast physical sizes. Thus Tycho's "principal argument" must still have been in play. Note that, based on Cassini's calculations, a star of apparent diameter of one second, approximately a fifth to a sixth that of Sirius, would still have a physical diameter of a fifth to a sixth that of Sirius (16 to 20 solar diameters) were it at the same distance. This is just a bit larger than the value Peter Crüger calculated almost a century earlier for a one arcsecond star (recall from chapter 4).

A century after Halley's comments, the English astronomer George Biddell Airy (1801–1892) developed a full theoretical explanation for the spurious disks of stars. It explained both the disks and why they vary in sizes for different stars. The explanation is that a star's spurious disk (often referred to today as its "Airy Disk") is a manifestation of the wave nature of light. Light waves entering a telescope from a tiny source of light undergo a phenomenon called *diffraction* (a phenomenon discovered, interestingly enough, by Francesco Maria Grimaldi) which produces a *diffraction pattern* consisting of a central maximum or disk and an indefinite number of surrounding rings of light whose brightnesses decrease continually according to fixed mathematical ratios, as seen in figure 10.2.[25] The rings extend outward indefinitely, but only a few are bright enough to be seen, and only in the case of brighter stars.

The ratio of the brightness of the central disk to that of the rings is also fixed, which has an important consequence for observations of stars. Recall from the previous chapter that Riccioli spoke of the telescope "exposing the disks of the stars and scraping off the adventitious ringlets of the rays" (denudanti discos stellarum, et abradenti cincinnos radiorum

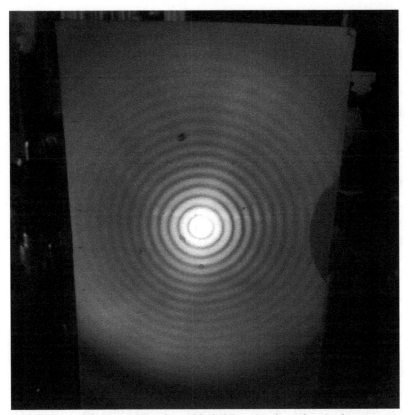

Figure 10.2.   Diffraction pattern formed by light passing through a circular aperture, showing the central spurious disk (or "Airy Disk") and the rings of decreasing brightness. Image credit: Wikimedia Commons. Available at http://en.wikipedia.org /wiki/File:Airy_disk_created_by_laser_beam_through_pinhole.jpg.

aduentitios). The ringlets—the *cincinnos*, that is, the diffraction pattern rings, which Riccioli interpreted as being part of the overall adventitious rays phenomenon—can indeed be "scraped off." This is done by reducing the amount of light entering the telescope, and thus reducing the brightness of the diffraction pattern, to the point where even the brightest (innermost) ring cannot be seen, and all that is visible is the central, entirely spurious, disk. When this is done using one of the brightest stars, the spurious disks of lesser stars seen through that telescope will appear smaller. This is because the brightness of the central maximum in the diffraction pattern does not fall off abruptly, but rather declines gradually, and so for fainter and fainter stars, less and less of their spurious disks' central maxima are visible.

One way to reduce the amount of light entering the telescope is to reduce the aperture of the telescope. Riccioli discusses doing just this. He describes how the object lens of a telescope must be masked with foil, or some other thin material, which has a hole smaller than the diameter of the lens. With such an apparatus, he says, a perfect, round disk will be seen, which can be compared with the disks of Jupiter or Saturn. In fact, because there is a fixed ratio between the brightness of the central disk and the brightness of the first (innermost) diffraction ring, various observers who might try Riccioli's technique, reducing the aperture until the last of the *cincinnos* is scraped off (that is, until the amount of light entering the aperture is sufficiently reduced to the point where the brightest, innermost ring in the diffraction pattern is no longer visible), will obtain reasonably consistent and reproducible results.[26] The nature of diffraction, combined with the theory that a telescope removes adventitious rays to show the true disk of a star, combine to produce *a consistent, yet entirely spurious* result. It is no surprise that even in 1720 Halley was having to debunk the notion that these disks were real (and thus that the stars were gargantuan bodies). But since these disks are spurious—produced within the telescope—they vanish the moment the light that the telescope is using to produce them is cut off, such as when the Moon passes in front of a star. Hence the discovery by Horrocks of stars disappearing "as if in the blink of an eye."

Another way to reduce the light entering a telescope is to filter it through a material such as smoked glass. This was the approach Huygens used. Unlike aperture reduction, this method does not appear to yield a clear disk, just as he argued in the latter part of the seventeenth century.[27]

Therefore, eventually, through the methods of Horrocks and Huygens, astronomers came to understand that the telescopic disks of stars are entirely spurious and a result of diffraction. According to Airy's diffraction theory, the overall size of the diffraction pattern with its central "Airy disk" varies inversely with the aperture of the instrument used to observe them. The human eye has an aperture of roughly a quarter that of the ¾-inch aperture of the telescope Flamsteed used when he measured Sirius to have a diameter of 15 seconds. Thus the Airy disk produced by a human eye when observing Sirius is roughly four times larger than that produced by Flamsteed's telescope: 60 seconds (one minute) in diameter. This, combined with distortions from the Earth's atmosphere, explains the star diameters Tycho Brahe determined with his nontelescopic instruments.[28] Today, no astronomers use a telescope with an aperture as small as ¾ inch. Even the smallest modern telescope, such as might be given as a gift to a child, has an aperture triple that of Flamsteed's. Thus even the smallest modern telescope, when used to observe Sirius, produces a spurious disk roughly a third what Flamsteed measured. Telescopes used by even casual amateur astronomers have apertures ten times that of Flamsteed's telescope; the spurious disks that such telescopes produce when used to observe Sirius are ten times smaller than what Flamsteed saw. At such a small scale, distortions from the atmosphere often create other spurious effects, so that seen through modern telescopes stars appear more like sparkling points than disks. In short, no modern astronomer, not even of the amateur child variety, uses a telescope with an aperture as small as Flamsteed's, and thus no modern astronomer sees stars the way Flamsteed (and Marius and Galileo and Hortensius and Riccioli) saw them. All astronomers today understand that "stars are points," even though the nature of stellar telescopic images was still a matter of debate as late as the mid-nineteenth century.[29]

Not until the mid-nineteenth century were final, complete solutions to Riccioli's anti-Copernican arguments developed—in the forms of a full understanding of the spurious appearance of the stars and of the Coriolis effect (including the advent of the Foucault pendulum). In the nineteenth century, annual parallax was also finally observed in the stars. This was accomplished in the 1830s, using the double star method of differential parallax suggested by Galileo and a telescope of sufficient power and precision to actually detect the effect. The effect of annual parallax in even the closest stars measures well less than one second of arc, indicating that those stars are distant by hundreds of thousands of solar distances.

We know today that stars are other Suns, so stupendously far away that they appear to the eye and to the telescope as mere points of light, as anyone who looks up, or who visits an observatory, can see. We know today that we live on an Earth that rotates diurnally, as anyone who visits a science museum with a Foucault pendulum can see. But this sure, demonstrable knowledge has only been available since the mid-nineteenth century.

The nineteenth century also saw the end of the theological arguments that Francesco Ingoli had presented to Galileo in his 1616 essay. Recall from chapter 5 that those theological arguments seemed to have two main thrusts, one being to convey some sense of Bellarmine's views on the matter, the other being to emphasize how the words of scripture fit a geocentric universe. Ingoli noted in his essay that "in explaining Sacred Writings the rule is to always save the literal sense, when it can be done, as in our case." Bellarmine had emphasized this point in April 1615, in a letter where he discussed the importance of the traditional, literal interpretation of scriptural verses that indicate that the Earth is at rest and the Sun moves, but where he also stated that finding a true demonstration of the reality of the Copernican system would require proceeding with great care in reinterpreting the scriptures, rather than claiming as false something that had been demonstrated true. As we saw in chapter 5, the French Jesuit Honoré Fabri was apparently the first to publish this sentiment. In 1661 he wrote that the Church would understand the scriptural passages that speak of the Sun's motion in a literal sense, so long as no demonstration to the contrary was available, and that if some demonstration of the validity of the Copernican hypothesis were to be found, the Church could declare that those passages are among those that are to be understood in a figurative sense.

Solid evidence for the Copernican hypothesis began to be found even before the nineteenth-century discoveries that finally answered Riccioli's main arguments. These included the development, in the late seventeenth century, of Isaac Newton's physics. This could explain Earth's motion and a variety of other phenomena in a Copernican universe, but not in a Tychonic one. The Copernican hypothesis was also supported by the detection, in 1728 by the English astronomer James Bradley, of an effect of Earth's motion on the light coming from stars—an effect known as *stellar aberration*.[30] As Rome began to reconsider the prohibitions against the Copernican system, the Inquisition consultant and Jesuit Pietro Lazzari argued in 1757 for lifting such prohibitions. Lazzari said that the evidence was indeed against Copernicus in 1616:

Thus, one can say that at that time there were good reasons or motives for [the Copernican system] being prudently prescribed. I consider three of these reasons for prescribing it. Firstly, this opinion of the earth's motion was new and was rejected and branded with serious objections by most excellent astronomers and physicists. Secondly, it was deemed to be contrary to Scripture when taken in the proper and literal sense; and this was conceded even by the defenders of that opinion. Thirdly, no strong reason or demonstration was advanced to oblige or counsel us to so disregard Scripture and support this opinion. One cannot deny that the collection of these reasons was a good and strong motive for adding that clause to the *Index*. And although decrees usually do not include the reasons, still in the one issued then, one has enough to understand that these were the motives of the prohibition.[31]

But while astronomers and physicists may have raised objections to the hypothesis in 1616, and the Copernicans may not have had any demonstrations in favor of the hypothesis at that time, the situation in 1757 is different, Lazzari says. In 1757 the Copernican system is widely accepted, he says. It is discussed even in Catholic academic journals. A book published in 1756 in Pisa by a Barnabite priest, which opens with a discussion of "phenomena that have become known everywhere in our century . . . [and that] confirm the most elegant and celebrated opinions of Galileo," was published with the usual approvals. It also received the imprimatur of the general of the Barnabite order, based on the reports of two of his theologians.[32] Lazzari continues:

> There is no need to speak of other books since it is clear and known to anyone of average education that nowadays the prevalent opinion among the most competent astronomers and physicists is that the earth moves around the sun.[33]

Today the Copernicans have strong support for their system, Lazzari writes. He says,

> Whereas formerly the Copernicans advanced in favor of their opinion reasons that were not very convincing, now they advance some that may be called demonstrations; thus these would suffice to persuade us.[34]

This was true even if that meant scripture had to be interpreted in a difficult way (which he notes, it does not, since it is easy enough to reconcile
scripture with the Copernican system simply by saying that scripture is
speaking of the motion of the Sun and stability of the Earth as we see
them). Lazzari continues:

> There is no doubt that if at the time of Galileo some demonstration
> had been adduced, one would not have proceeded to the prohibition
> of his book and of the Copernican system. There is about this an ex-
> plicit testimony and a letter written by Father Honoré Fabri published
> in the *Transactions* of June 1665. Wolff reprinted it in his *Elements of
> Astronomy*: "Rewriting what Father Fabri once publicly asserted, the
> Church does not operate against the evidence, but will declare that the
> system of a moving earth is not contrary to Scripture as soon as a
> demonstration of it is put forth."[35]

What are these demonstrations? According to Lazzari, Bradley's discovery of the aberration of starlight and the system of Newtonian physics.
For once the Newtonian system was established, scientists could seek out

> all the most minute and individual consequences that derived from it;
> these consequences embrace a prodigious quantity of things which
> are so complex that by means of pure observation it had not been pos-
> sible to infer their laws and to calculate and predict them; only in that
> system were they seen to come about as a legitimate consequence.[36]

These include, Lazzari writes, the many aberrations in lunar motion, the
laws of the tides, and the motions of comets, among other things. Thus
the Copernicans now have powerful demonstrations that theirs is the true
system of the world.[37]

The Dominican friar Maurizio Benedetto Olivieri (1769–1845), another Inquisition consultant, would put forth these same sorts of arguments in the 1820s. At this time the Copernican system was again being
discussed in Rome in the face of still further discoveries, including Guglielmini's apparent detection of Earth's rotation via his falling bodies experiments and Guiseppe Calandrelli's (1749–1827) claimed detection of an-

nual parallax in stars.[38] Challenging an opponent in this debate, Olivieri, who among other things notes that the aberration of starlight has been verified for a century,[39] cites the importance of evidence for the Copernican system (or lack thereof) in the condemnation of 1616:

> [Is] it not true that the doctrine declared "heretical or at least erroneous in the Faith" was also "philosophically false and absurd"? You do not deny, nor could you or anyone else deny, that the theological consultants said so. The same identical doctrine was judged "philosophically false and absurd," and then "formally heretical or at least erroneous in the Faith." Implicit here is a syllogism, with major and minor premises, and it is accepted by you, me, and all. But the doctrine of modern astronomers . . . can no longer be called "philosophically false and absurd [i.e., scientifically untenable]" by anyone.[40]

The general prohibition against books teaching the Copernican system was dropped by Rome in 1758, and the specific prohibition against Copernicus's *On the Revolutions* and Galileo's *Dialogue* in 1835.[41] Bellarmine's true demonstration had been provided. The absolute literal sense of scriptural passages about the motion of the Sun and the immobility of Earth could not be retained. Theological arguments were conditional on the scientific facts. Where scripture obviously disagrees with science or mathematics, it is understood that the literal sense is not to be retained. In 1 Kings 7:23 a round pool of water is said to have a circumference that is three times its diameter, meaning that pi must be equal to three. Revelation 7:1 describes the world as having four corners. Mark 4:31 declares the mustard seed to be the smallest seed on Earth. Genesis 22:17 suggests that the number of stars visible in the sky and the number of grains of sand seen on the shore are comparable.[42] These passages are not read as describing the world with scientific precision. Science obviously shows that, for example, the world does not have four corners. Mathematics is clear that the circumference of a circle is not precisely three times its diameter. As solid evidence was now in hand for the Copernican system, those passages that speak of a moving Sun would simply join these.

But during the period from the late sixteenth century to the end of the seventeenth, solid scientific arguments by Tycho Brahe and Giovanni

Battista Riccioli supported the idea of a moving Sun. The Copernican system with its giant warrior stars looked foolish and absurd—not just in the eyes of the consultants for the Roman Inquisition, but in the eyes of the leading astronomer of the time, Tycho Brahe. Thus it could be considered "heretical," as it contradicted the most plain reading of scripture when there was no reason for any other reading.

The strength of anti-Copernican science has been forgotten by history, and anti-Copernicans like Giovanni Battista Riccioli have come to be characterized as the minions of villainous superstition—the sort of people who based their opinions on nothing other than authority, scripture, and decrees from Rome. Likewise, the weakness of Copernican arguments has been forgotten by history, and Copernicans have come to be characterized as the allies of brave reason. But arguably anti-Copernican science, supporting the hybrid geocentric Tychonic system, reached its zenith with the *New Almagest*, a book written by an Italian Jesuit geocentrist who, with a cadre of priests, employed experimentation and the telescope against an "absurd" and "fraudulent" Copernican hypothesis whose backers invoked Divine Omnipotence to solve its scientific problems. In 1651, decades after the advent of the telescope, one could ask which hypothesis could be asserted as being true, according to pure reason—that which supposes the motion of the Earth or that which supposes the immobility of the Earth?—and one could answer quite reasonably and confidently:

> Reasoning and intrinsic arguments alone considered, and every authority set aside; the hypothesis supposing the immobility or quiet of the Earth absolutely must be asserted as true; and the hypothesis that bestows to the Earth motion (either solely diurnal, or diurnal and annual) absolutely must be asserted as false and disagreeing with physical and indeed physico-mathematical demonstrations.[43]

Appendix A

A Rendition into English and a Technical Discussion
of Francesco Ingoli's 1616 Essay to Galileo

What follows in part 1 of this appendix is an effort to render Ingoli's Latin into a form relatively accessible to the modern reader. While we (Christina Graney provided invaluable assistance in translating this work) strove to produce a faithful rendering of Ingoli's work, there were places where we felt the need to break the essay into additional paragraphs, or to paraphrase somewhat. Also, we have frequently divided lengthy Latin sentences into multiple shorter English sentences. Our paragraph breaks are indicated by indentations; Ingoli's original paragraphs are indicated by line spaces. For the reader who wishes to see the essay as Ingoli wrote it, his original Latin is included in part 2 of this appendix. The Latin is taken from Favaro 1890–1909, 5:403–12. Lastly, part 3 of this appendix consists of a technical discussion of the essay.

PART 1
English Rendition of Ingoli's 1616 Essay to Galileo

FRANCESCO INGOLI
RAVENNA

Disputation concerning the location and rest of Earth
against the Copernican system

TO THE LEARNED FLORENTINE MATHEMATICIAN

D. GALILEO GALILEI

PUBLIC PROFESSOR OF MATHEMATICS,
FORMERLY IN THE GYMNASIA TO THE PADUANS,
BUT NOW THE PHILOSOPHER AND CHIEF MATHEMATICIAN
OF THE MOST SERENE GRAND DUKE OF ETRURIA.

Preface

Among the many disputations that have come before the Very Distinguished and Most Reverend D. Lorenzo Magalotti (a committed man of prudence and letters in the Roman Curia), a particular and singular one has been that concerning the situation and motion of the Earth, as in the position of Copernicus. In this disputation you, learned man, certainly are the defender of Copernicus, offering arguments to solve those of Ptolemy, and endeavoring to confirm the Copernican system. But I have been given the role of defending the other side, bringing forth arguments to sustain the hypothesis of the old mathematicians, and to tear down the Copernican assumption. Now finally, things have come to the point for the truth to be demonstrated about your promised experimental solution to the argument of Ptolemy, and for the argument proposed by me concerning parallax to be presented in writing, in order that you might be able to produce a timely solution of it.

I have agreed to my role extremely willingly. I am always thankful to be among the most learned men, such as yourself, engaged in the best sort of debate. Indeed I am honored by such debates, and in them I often learn some things.

And so now, back home, I need to fulfill my duties. But in thinking of this it occurred to me to contact you, because you might willingly hear each and every argument in this disputation which might adduce reasoning against Copernicus. Thus in order that the truth of the thing might be investigated more easily, I have resolved to write not only the argument concerning parallax, but others likewise (although not all), which can be made against the Copernican system and the motions of the Earth devised from it. If you too would think it worthwhile to write to satisfy these arguments, it will be most pleasing to me, and I will be most grateful to you.

Chapter 1: The Structure of This Letter

I will follow this method in this disputation: I shall discuss first against the positions of the Earth and the Sun, specified by Copernicus in his system, and second against the motions of the Earth and the immobility of the Sun. I will generate three types of arguments: mathematical, physical, and theological.

Chapter 2: Mathematical Arguments against the Copernican Position of the Earth

Copernicus proposes the Sun to be in the center of the universe, while the Earth is in a circle between the orbs of Venus and Mars.

Against a position of this sort, first, I present the argument concerning parallax. For if the Sun were in the center of the universe, it would admit greater parallax than the Moon. But the consequent is false, therefore, so is the antecedent. The logical consequence is demonstrated: because by how much bodies are more remote from the prime mover (on which their locations are noted by astronomers), by that much they admit greater parallax, while the aspect of difference corresponds to both the theories and the tables (in which the parallax of the apogee Sun is noted smaller, because at that time it is closer to the prime mover, but the parallax of the perigee Sun is noted larger, because it is more remote). Yet the Sun (according to Copernicus) is more remote from the prime mover than the Moon; because she is outside the center, and he is truly in the center, and the center is a place more remote from the periphery. Therefore the Sun will admit greater parallax. The falsehood of the consequent is certainly easily demonstrated: for from observations it is manifest, the greater parallax of the Sun to be 2 minutes 58 seconds, of the Lunar party to be truly 1.6 minutes 21 seconds, as Magini[1] from Reinhold[2] has stated, in *Theories*, book 2, chapter 20, at the end. From which observations it is evident, the parallax of the Sun not to be greater than the parallax of the Moon, but the latter to surpass the former by far, as the number 22 without doubt surpasses unity.

Nor does it satisfy if it is said, therefore, that the Moon has greater parallax because it is closer to us (since it stands apart from Earth by only 52 and 17/60 up to 65 and 30/60 terrestrial semidiameters from the fourth limit to the first,[3] as Magini notes from Copernicus, second book chapter 24 of *Theories*; and the Sun stands apart by 1179 terrestrial semidiameters). First, because if this solution would prevail, it would necessitate that the distances of the luminaries have whatever proportion to each other that their parallaxes have, since the parallaxes depend on the distances. But we do not see this because the distances are in the proportion of 18 to 1, as Magini notes from Copernicus, but the parallaxes are 22 to 1, as said previously. Therefore, the solution does not prevail. Second, because the quantity of parallax not only determines the distances of bodies from us seen in terms of height, but also the distances from the eighth orb [i.e., the orb of the stars], whereby parallaxes are noted. And so since the Sun (according to the observations of Copernicus) stands apart from the starry heaven by 1244 terrestrial semidiameters[4] more than the Moon, when it is in opposition of the Sun, it does not seem to me to be possible that the parallax of the Sun may be 1/22 of the parallax of the Moon.

The second argument is from Sacrobosco,[5] in *Sphere*, chapter 6. This says Earth is in the center of the eighth orb, because the stars appear to us to have the same size regardless of their elevation above the horizon. This would not be true if Earth did not occupy the center. This is proven from the definition of the circle (only lines that lead from the center to the circumference are equal to each other) and from the rule of view (those things that appear larger to us are closer because they subtend a greater angle; those that appear smaller are more remote because they subtend a smaller angle).

The third argument is by Ptolemy, book 1, chapter 5, of the *Almagest*, saying, the Earth is in the center of the universe because of the reason that, wherever a man may stand, he always sees half of the sky (that is, 180 degrees). It would not be such if Earth were outside the center. But that half of the sky may be visible anywhere is evident not only from opposed fixed stars (obviously from the Eye of the Bull and the Heart of the Scorpion, one of which rises while the other sets); but also from the certain observation of 90 degrees, which can be made while the Sun is in the points of Aries or Libra, if the meridian elevation of the equator is noted, and the intervening distance from that up to the pole may be observed, and finally since by this the eastern part and the western part of the vertical circle may be measured.[6] That truly half of the sky would not be visible if Earth did not occupy the center is evident from the definition of the semicircle: indeed only the diameter, which always crosses through the center of the circle, divides the circle itself into two equal semicircles.

Nor does the solution entirely satisfy by which is said: the diameter of the circle of the orbit of Earth in comparison to the vast distance of the eighth orb from us is made so small that it may subtend only 2 minutes on the eighth orb itself.[7] For if the Earth is to be insensible in magnitude with regard to the starry orb, it is necessary that she may be distant from it by fourteen thousand of her own semidiameters (according to Tycho's pleasure, as is seen in his book of *Astronomical Letters* in response to the letters of Rothmann, page 188). It is likewise necessary that the circle of the orbit of Earth (the semidiameter of which is 1179 terrestrial semidiameters, if we trust Magini, who writes, in book 2, chapter 24 of *Theories*, the distance of the apogee Sun from Earth, to be of such size according to the Copernican observations) may be distant from the eighth sphere by 14,000 of its own semidiameters, which make 16,506,000 terrestrial semidiameters. Such a truly great distance not only shows the universe to be asymmetrical, but also clearly proves either the fixed stars to be unable to work in these regions below, on account of the excessive distance of them (which can be proven from that which happens in the case of the Sun; for we experience the power of it to be made so dull in winter that we may distinctly perceive great cold, on account of the distance of it from the Zenith of our head—certainly nothing in comparison to the distance of Earth from the eighth orb), or the fixed stars to be of such size, as they may surpass or equal the size of the orbit circle of the Earth itself, whose semidiameter is, as we have said, 1179 terrestrial semidiameters. This can be proven from the apparent size of the solar body. For if the Sun is seen by us to have a diameter of 32 minutes at a distance from the Earth of 1179 terrestrial semidiameters, how great ought to be the size of the fixed stars, which are distant from Earth by 16,506,000 terrestrial semidiameters, when they may appear to us to be 3 minutes, following the old opinion, or even 2 minutes, following your pleasure? And so by reason of these I think the arguments of Sacrobosco and Ptolemy cannot be solved through the assumption that the diameter of the orbit of the Earth may subtend only 2 minutes in the firmament of heaven.[8]

The fourth argument is Tycho's, from the *Astronomical Letters*, page 209, where he shows by the most reliable experiments that the eccentricities of Mars and Venus are discovered to differ greatly from what Copernicus noted; and likewise the apogee of Venus is found not to be immobile (as the same Copernicus has affirmed), but to be moved within the sphere of the fixed stars. Because of these, great doubt is cast upon the system of Copernicus, since it was established to explain such phenomena and it does so with minimal satisfaction.

Chapter 3: Physical Arguments

Two arguments seem to me to show the Earth to be in the center of the universe. One is taken from the order of the universe itself. For we see in the arrangement of simple bodies, the denser and heavier to occupy the lower place, as is well known concerning earth with respect to water and concerning water with respect to air. Earth is denser and heavier than the solar body, and the lower place in the universe is without doubt the center. Thus the Earth, and not the Sun, occupies the middle or center of the universe.

But if the first part of the minor proposition of this argument were to be denied, it can be proven. First, by the authority of the Philosopher [Aristotle] and all the Peripatetics, who say the heavenly bodies have no heaviness. Second, by logical reasoning even, for the opposite proposition is this: the Sun is a body denser and heavier than Earth. At first thought this is seen to be false, when we may see all things which have luminosity to be lighter and less dense, as is well known concerning fire, and concerning those things that are released from it.

If truly the second part were to be denied, it can be proven by the authority of the philosophers, saying the position of the center of the universe to be a down location, or lower, and the circumference of the same to be an up location, or higher. And it can be proven by reasoning, because in the globe of the Earth itself we designate as the higher parts those that are located toward the periphery of it, and the lower those that are located within the circumference and toward the center, while we so designate the center itself to be the lowest part of the Earth. The center therefore is the lower location in the universe.

The second argument is that taken from bits of the Earth itself. For in unsifted wheat we see lumps of dirt in the wheat, drawn to the center of a sieve by the sieve's circular motion. The same happens in the case of heavier bits of gravel, when they are stirred about in any round vessel. By such an experiment many philosophers have maintained the Earth to stand in the middle of the universe, because that way it is dispossessed from the motions of the heavens. But if this happens to bits of the Earth, likewise it must be said to happen to the whole of the Earth, since the argument from part to whole may hold in the homogeneous. Also Rothmann, defending Copernicus in his letter which is in Tycho's book of *Astronomical Letters*, page 185, uses this genus of argument from parts of Earth to the whole Earth.

Chapter 4: Theological Arguments

Finally, as I shall conclude the first part of this disputation that the Earth and not the Sun is in the middle of the universe, two other arguments, taken from Sacred Scripture and from the doctrine of the theologians, stand out to me. One of these is taken from chapter 1 of Genesis. Ponder the words: "And God said: Let there be lights made in the firmament of heaven."[9] For in the Hebrew text the name רקיע *rakia* signifies expanse or extent or extension; one of these words may be considered in the place of the word *firmament*, as Saint Pagninus recommends in the *Thesaurus of the Holy Language* in regard to the root *raka*. And a meaning of this sort is in no way appropriate to a center, the nature of a center itself being repugnant to an extension or, as thus I may say, to an expansion. It may, however, be suitably appropriate to the circumference of the heaven, which is in a certain way extended and expanded above the center (whence, the appropriate metaphor in Psalm 103:2 referring to God "Who stretchest out the sky like a pavilion").[10] The greater fact needing to be said is that God said, "Let lights be made in the firmament of heaven"—not in the center to illuminate, but in the expanse or extent itself of heaven. This argument is strengthened because the word *Fiant* [third person plural verb], which God has said, considers equally Sun and Moon, since in the text it is said, "Fiant luminaria in firmamento coeli"; whence as the Moon is not in the center, but in the expanse of the heaven, thus also the Sun ought to be in the expanse, and not in the center.

Another argument is from the doctrine of theologians. This holds mainly on account of the reasoning that Hell, that is, the place of the demons and of the damned, is in the center of Earth, because, since Heaven may be the place of the angels and the blessed, it behooves the place of the demons and the damned to be in the most remote place from Heaven, which is the center of Earth. Thus Hell and Heaven are appointed places most distant from each other, as Psalm 138 has said: "If I ascend into heaven, thou art there: if I descend into hell, thou art present."[11] Also Isaiah 14, where is said to the king of Babylon, and to the devil in the shape of him: "And thou saidst in thy heart: I will ascend into heaven. . . . But yet thou shalt be brought down to hell, into the depth of the pit."[12] Read the Most Illustrious Cardinal Bellarmine, *De Christo*, book 4, chapter 10, and *De Purgatorio*, book 2, chapter 6. And so since Hell is in the center of Earth, and Hell ought to be the most remote place from Heaven, Earth is admitted to be in the middle of the universe, which is the most remote place from heaven. This marks the end of the first part of this disputation.

Chapter 5: Mathematical Arguments against
the Copernican Motion of the Earth

Many things can be presented against the diurnal motion of Earth. Some of these Tycho directs against Rothmann, the defender of the opinion of Copernicus, in two astronomical letters in the book of *Astronomical Letters*, pages 167 and 188. One is concerning the fall of a lead ball perpendicularly from the highest tower (under the pretense of no resistance from the attending air, when nevertheless it ought to be the contrary), because Earth by diurnal motion, even in the northern parallels of Germany, would be moved 150 greater paces in a small second of time. A similar one is concerning the bombards discharged from the east into the west and from the north into the south, particularly concerning those discharged near the poles, where the movement of Earth is slowest. For, given the diurnal motion of Earth, the most apparent differences would be observed, whereas nevertheless no differences are observed.

Truly many more things can be presented against the annual motion. About these see Tycho above. I shall bring up only four reasons. The first of these is taken from the risings and settings of the fixed stars. For if Earth is moved by annual motion, it necessitates the latitudes of the fixed stars rising and setting to be varied sensibly over eight or ten days. But the consequent is false, and therefore so is the antecedent. The falsehood of the consequent is recognized because the aforementioned latitudes do not vary notably unless over fifty or sixty years. The consequent truly is proven, because it is necessary that—when the Earth together with the horizon may be moved within the zodiac, and thus moved from south to north and north to south, and moved sensibly in eight or ten days, and the fixed stars moved truly insensibly on account of their so slow motion within the zodiac (they are actually immobile, following Copernicus)—the fixed stars themselves would notably vary their latitudes of rising and setting in the space of eight or ten days.

The second is from the polar altitudes of places. For if Earth is moved by annual motion, it requires the polar altitudes of places to be changed. But the consequent is false, and therefore, so is the antecedent. The falsehood of the consequent is well known—the polar altitudes do not change. That they must be changed is proven because, when Earth may be carried from north to south and south to north by means of annual motion, the places existing on it are also simultaneously thus carried. But that proposal entirely changes the polar altitudes. Indeed, just as the man who makes a journey from south to north or north to south observes the altitude of the pole to change, likewise if the place he is on itself moves from south to north or north to south he will also observe the altitude of the pole to change.[13]

The third is from the inequality of conventional days [the varying length of daylight[14]]. For, although they may appear to be wholly consistent with observations, given the annual motion of the Earth and the stillness of the Sun, because the rectitude or obliqueness of the horizon always appears the same (since the horizons may be presupposed to be moved together with the Earth), nevertheless it will be seen not so upon close inspection. Because, since Earth may be carried by means of the annual motion from north to south and south to north, it is necessary that the zenith above our head may be similarly carried. By reason of consequence, at one time it may approach the [celestial] equator and at another time it may move away. But from the change of the zenith the rectitude and obliqueness of the horizon is certain to be changed, which principally produces the inequality of the days. From which it follows, that two things would need to be noted for marking the inequality of the days—the differences of the annual [solar] motion, effecting the parallels of the conventional days, which happens according to the posited motion of the Sun and stillness of the Earth; and those differences that might produce changes to the obliqueness of the horizon, given the motion of the Earth and the stillness of the Sun, especially around the northernmost habitations, where the variations of the days are most sensible. Yet this does not happen, since considering only the first differences satisfies the observations.

It hinders not, that the horizons may be carried together with Earth without change of them, because this might be true with respect to the diurnal motion of Earth but not with respect to the annual. For with respect to this, even in the case that they may be transported with Earth, still they are changed with respect to obliqueness and rectitude, on account of the necessary change of the zenith, as was said.

The fourth is from Tycho in the book *Astronomical Letters*, page 149, where he asserts that the movement of comets across the sky when in opposition of the Sun does not at all reflect an annual motion of the Earth. This they ought to do because, unlike in the fixed stars, it is not the case that a reflection of annual motion should vanish, since the aforementioned comets may not have that most great distance from Earth of the fixed stars.

Finally Tycho in the aforementioned *Letters* presents three arguments against the Earth's third motion. First, the third motion is necessarily removed if the annual motion is eliminated. Second, it is not possible that the axis of Earth may truly gyrate annually, in a manner contrary to but corresponding to the annual motion, such that nevertheless it may be seen to rest. Third, it is not possible to be granted in a single and simple body for the axis and center to be moved by a double and opposite motion (if the diurnal motion may in addition be added to that, more difficulty is produced).

Chapter 6: Physical Arguments against the Motion of the Earth

Many things that are produced by the philosophers and mathematicians on behalf of a resting Earth (and especially by Tycho in the aforementioned *Letters*) might adduce physical arguments against the motion of Earth. But I shall put forward only three common ones. One of these is from the nature of heavy and light bodies. For in general we see heavy bodies to be less apt to motion than light or not heavy, which indeed can become known immediately by considering not only simple natural bodies but also mixed, and these not only in turn according to motion that is caused by an intrinsic principle but also in turn according to motion that takes place by an extrinsic principle. Again we see nature adapt materials to forms thus, as according to the efficiency of the forms themselves we may notice the amazing aptitude of the materials themselves. This occurs first because, as the Philosopher says, in *Physics*, book 2, nature acts on account of an end, then because materials are like the instruments of the forms according to the acting. And so since Earth may be the heaviest of all bodies subject to our knowledge, it is by no means proper to say that nature bestowed so many motions to it, and the diurnal motion especially—so swift that in one minute of time Earth would traverse nearly nineteen miles, as Tycho says in the *Astronomical Letters*, page 190.

Another is that which is taken from the physical proposition that to each one natural body belongs only one natural motion (which might be easily proven to be true by induction, even were it not supported by the most excellent philosopher). Thus since the natural motion of Earth may be toward the center, motion around the center will not be able to be natural to it, and much less will so many motions (and all not to the middle) be able to be natural to it. If then those motions of Copernicus are not natural to Earth, how can it happen that Earth, a natural body, may be moved by those for all time? For it is not of nature to be driven contrary to nature.

The third is from a certain incongruence: because of course to all bright parts of the heaven, certainly to the planets, Copernicus has attributed motion. Yet to the Sun (of all parts of the heaven the most outstanding and bright) he denies motion, while to Earth (the dark and dense body), he assigns motion. Nature, most discrete in all of its works, ought not to do this.

Chapter 7: Theological Arguments against the Motion of Earth

Endless theological arguments from Sacred Scripture and from the authority of the Fathers and of the Scholastic theologians might be able to be proposed against the motion of Earth, but I shall adduce only two, which seem to me to be more substantial.

One is from Joshua, chapter 10, where on the prayers of Joshua, Scripture says: "So the Sun stood still in the midst of heaven, and hasted not to go down the space of one day. There was not before nor after so long a day, the Lord obeying the voice of a man."[15] The responses to this that are produced, namely, that Scripture may speak following our manner of understanding, do not satisfy: first because in explaining Sacred Writings the rule is to always save the literal sense, when it can be done, as in our case [through the Tychonic system]; next because all the Fathers unanimously explain this passage to mean the Sun, which moved, truly stood still on the prayers of Joshua. Indeed the Tridentine Synod (fourth session, in the doctrine concerning the publication and use of Sacred Books, § *Praeterea*) is averse to any interpretation that is against the unanimous consensus of the Fathers. And granted that the Holy Synod [Council of Trent] may speak in the matter of morals and of the Holy Faith, nevertheless it is not able to be denied, but that the interpretation of the Holy Scripture against the consensus of the Fathers may displease those Holy Fathers.[16]

Another is by authority of the Church: for in the hymn at vespers of the third day [Tuesday] thus she sings:

> Earth's mighty Maker, whose command
> raised from the sea the solid land;
> and drove each billowy heap away,
> and bade the earth stand firm for aye.[17]

The origin of this sort of argument is not trivial. Such is seen in the writings of Cardinal Bellarmine. In many passages he refutes numerous errors by hymns, songs, and prayers of the Church, which are found in breviaries.

These complete this disputation. Let it be your choice to respond to this either entirely or in part—clearly at least to the mathematical and physical arguments, and not to all even of these, but to the more weighty ones. For I have written this not toward attacking your erudition and doctrine (most notable to me and to all men both inside the Roman Curia and outside), but for the investigation of the truth, which you profess yourself always to search for by all strength, and in fact so suits a mathematical talent.

THE END

PART 2
Ingoli's 1616 Essay to Galileo

FRANCISCI INGOLI

RAVENNATIS

De situ et quiete terrae contra Copernici
systema disputatio

AD DOCTISSIMUM MATHEMATICUM

D. GALILAEUM GALILAEUM

FLORENTINUM

PUBLICUM PROFESSOREM MATHEMATICARUM
OLIM IN GYMNASIO PATAVINO, NUNC AUTEM
PHILOSOPHUM ET MATHEMATICUM PRIMARIUM
SERENISSIMI MAGNI DUCIS ETRURIAE.

Prooemium

Inter multas disputationes quas apud Perillustrem et Reverendissimum D.
Laurentium Magalottum, virum ob prudentiam et litteras in Romana
Curia commendatum, habuimus, illa praecipua et singularis fuit de situ et
motu Terrae iuxta positionem Coperniceam. In qua tu quidem, vir doc-
tissime, Copernici partes defendendas assumens, plurima in medium
proferebas, quibus Ptolomaei argumenta solvere, et systema Copernici
comprobare, conabaris: ego autem, contra, veterum mathematicorum hypothesim sustinere, et Coperniceam assumptionem destruere, vario argumentandi genere pro viribus nitebar. Tandem, post multa, eo res devenit,
ut pro solutione argumenti Ptolomaei experimento, quod pollicebaris,
veritas probaretur, et argumentum de parallaxi a me propositum scripto
exhiberetur, ut maturius eius solutionem afferre posses. Annui perquam
libenter: nam cum viris doctissimis et in disputationibus modestissimis,
qualis tu es, agere gratissimum semper mihi fuit; aliquid enim plerunque
addisco, honoremque non minimum adipiscor. Domum itaque reversus,

promissa implere cogitavi: sed cum, inter cogitandum, mihi te dixisse occurrisset, quod in hac disputatione libentissime unumquemque audires, qui rationes contra Copernicum adduceret, ut sic facilius rei veritas investigaretur, deliberavi non solum de parallaxi argumentum scribere, sed alia quoque, licet non omnia, quae contra systema Coperniceum et Terrae motiones, ab eo excogitatas, fieri possunt: quibus si tu quoque scripto satisfacere dignaberis, gratissimum mihi erit, et plurimas tibi habeo gratias.

Ordo huius scriptionis.
Cap. Primum

Methodus autem in hac disputatione a me servanda erit huiusmodi. Primo, disseram contra situationem Terrae et Solis, quam ponit Copernicus in suo systemate; 2°, contra motus terreni orbis et Solis quietem: et in utroque capite triplici argumentorum genere, videlicet mathematico, physico et theologico.

Mathematica argumenta contra situm terrae Coperniceum.
Cap. 2m.

Proponit Copernicus, Solem esse in centro universi, Terram autem in circulo inter Veneris et Martis orbes.

Contra huiusmodi positionem, primum, obiicio argumentum de parallaxi. Nam si Sol esset in centro universi, maiorem admitteret parallaxim quam Luna: sed consequens est falsum: ergo, et antecedens. Consequentia probatur: quia corpora, quanto remotiora sunt a primo mobili, in quo eorum loca notantur ab astronomis, tanto maiorem admittunt parallaxim, ut ex diversitatis aspectus theoricis et tabulis constat, in quibus Solis apogaei parallaxis minor notatur, quia tunc vicinior est primo mobili, maior autem perigaei, quia remotior: sed Sol, iuxta Copernicum, est remotior a primo mobili quam Luna; quia haec est extra centrum, ille vero in centro, et centrum est remotior locus a peripheria: igitur Sol maiorem admittet parallaxim. Falsitas vero consequentis facillime probatur: nam ex observationibus manifestum est, Solis parallaxim maiorem esse 2′ 58″, Lunae vero partis 1.6′ 21″, ut ex Rehinoldo annotavit Maginus, Theoricorum lib. 2°, cap. 20 in fine; ex quibus observationibus liquet, non Solis parallaxim maiorem esse parallaxi Lunae, sed hanc illam longe superare, ut nimirum numerus 22 superat unitatem. Nec satisfacit si dicatur, ideo Lunam habere maiorem parallaxim, quia nobis vicinior est, cum distet a Terra semidiametris terrenis tantum 52.17′ usque ad 65.30′ a quarto limite ad primum, ut ex Copernico notat Maginus, Theoricorum lib. 2°, cap. 24; quibus Sol distat 1179. Primo: quia si haec solutio valeret, necessarium esset, ut quam proportionem habent luminum distantiae inter se, eandem haberent et parallaxes eorum, cum parallaxes a distantiis pendeant: hoc autem non videmus; quia distantiae se habent sicut 18 ad 1, ut Maginus notat ex Copernico ubi supra, parallaxes autem sicut 22 ad 1, ut dictum est: igitur solutio nihil valet. Secundo: quia parallaxis quantitatem non solum efficit distantia corporum visorum in sublimi a nobis, sed etiam distantia ab octavo orbe, ubi notantur parallaxes. Cum itaque Sol distet a caelo stellato plus quam Luna, quando est in Solis opposito, iuxta observationes Copernici, semidiametris terrenis 1244, non videtur mihi fieri posse ut parallaxis Solis sit 1/22 parallaxis Lunae.

Secundum argumentum est Sacrobusti, in Sphaera, cap. 6, dicentis, Terram, esse in centro octavi orbis, quia stellae in quacunque elevatione sint supra horizontem, eiusdem quantitatis nobis apparent: quod non esset, si Terra centrum non possideret. Quod probatur, tum ex diffinitione circuli; nam solae lineae quae a centro ad circumferentiam ducuntur, sunt inter se aequales: tum ex regula prospectivae, qua dicitur, quae maiora nobis apparent, viciniora esse, quia sub maiori angulo videntur, quae autem minora, remotiora, quia sub minori angulo conspiciuntur.

Tertium argumentum est Ptolomaei, lib. 1, cap. 5, Almagesti, dicentis, ideo Terram esse in centro mundi, quia, ubicunque existat homo, semper videt coeli medietatem, hoc est gradus 180; quod non esset si Terra esset extra centrum. Quod autem coeli medietas ubicunque conspiciatur, liquet non solum ex stellis fixis oppositis, nempe ex Oculo Tauri et Corde Scorpionis, quarum una oritur dum alia occidit; sed etiam ex certa observatione graduum 90, quae potest haberi dum Sol est in punctis Arietis vel Librae, si notetur elevatio aequatoris meridiana, et ab ea usque ad polum interiecta distantia observetur, et tandem cum hac mensuretur portio orientalis et portio occidua circuli verticalis. Quod vero medietas coeli non conspiceretur si Terra centrum non occuparet, constat ex diffinitione semicirculi: sola enim diameter, quae semper transit per circuli centrum, dividit ipsum circulum in duos semicirculos aequales. Nec solutio qua dicitur, diametrum circuli deferentis Terram in comparatione distantiae maximae octavi orbis a nobis adeo exiguam fieri, ut in ipso orbe octavo solum 20' subtendat, omnino satisfacit. Nam si Terra, ut insensibilis magnitudinis evadat respectu stellati orbis, necesse est ut distet semidiametrorum suarum quatuordecim millibus ab ipso, iuxta Tychonis placita, ut videre est in eius libro Epistolarum Astronomicarum in responsione litterarum Rothmani, pag. 188, oportebit quoque ut circulus deferens Terram (cuius semidiameter est semidiametrorum terrenarum 1179, si Magino credimus, qui distantiam Solis apogei a Terra, Theoricorum lib. 2, cap. 24, tantam esse scribit iuxta Coperniceas observationes) distet ab octava sphaera suis semidiametris m/14, quae faciunt semidiametros terrenas 16506000: quae distantia adeo magna non solum asymmetrum esse universum ostendit, sed etiam convincit, aut stellas fixas nihil operari posse in haec inferiora, propter nimiam earum distantiam (quod comprobari potest ex iis quae contingunt in Sole; nam experimur virtutem eius in hyeme, propter distantiam ipsius a Zenith capitis nostri, minimam certe in comparatione distantiae Terrae ab octavo orbe, adeo hebetem fieri ut frigus magnum persentiamus); aut stellas fixas tantae magnitudinis esse, ut superent aut aequent magnitudinem ipsius circuli deferentis Terram, cuius semidiameter est, ut diximus, 1179 semidiametrorum terrenarum: quod probari potest ex magnitudine apparenti corporis solaris; nam si Sol nobis videtur diametrum habere 32' in distantia a Terra semidiametrorum terrenarum 1179, quanta debebit esse magnitudo fixarum, quae distant a Terra semidiametris terrenis 16506000, ut nobis appareant esse 3', secundum antiquam opinionem, vel etiam 2', secundum placita tua? Ex his itaque puto, Sacrobusti et Ptolomaei argumenta minime solvi posse per assumptionem, quod diameter deferentis Terram subtendat solum 20' in coeli firmamento.

Quartum argumentum est Tychonis in dictis Epistolis Astronomicis, pag. 209, ubi probat certissimis experimentis, reperisse eccentricitates $\mars^{is}$ et $\venus^{is}$, notatas a Copernico, aliter longe se habere; sicut et apogeum $\venus^{is}$ non esse immobile, ut idem Copernicus affirmavit, sed sub fixarum sphera moveri: ex quibus valde dubium Copernici systema efficitur, cum phaenomenis, pro quibus salvandis ab eo sic constitutum est, minime satisfaciat.

Argumenta Physica.
Cap. 3m.

Terram esse in medio universi, duo argumenta mihi videntur ostendere; quorum alterum est, quod ab ordine ipsius universi desumitur. Nam in coordinatione corporum simplicium videmus, crassiora gravioraque inferiorem locum occupare, ut patet de terra respectu aquae et de aqua respectu aeris: Terra autem crassius et gravius corpus est corpore solari; et locus inferior in universo procul dubio est centrum: Terra igitur, et non Sol, medium sive centrum universi tenet.

Quod si negetur prima pars minoris propositionis huius argumenti, potest probari, primo, authoritate Philosophi et Peripateticorum omnium, dicentium corpora caelestia nullam habere gravitatem: 2°, ratione saltem logica; nam propositio opposita, hoc est, Sol est corpus crassius et gravius Terra, ipso primo animi conceptu videtur esse falsa, cum omnia quae habent lucem videamus esse rariora et leviora, ut patet de igne et de iis quae passa sunt ab eo.

Si vero negetur secunda pars, et philosophorum authoritatibus probari potest, dicentium positionem centri universi esse locum deorsum, circumferentiam vero eiusdem esse locum sursum, quod est idem ac si diceretur inferius et superius: et ratione; quia in ipso Terrae globo superiores partes dicimus quae ad peripheriam eius, inferiores vero quae infra circumferentiam et versus centrum, locantur, ita ut centrum ipsum infimam dicamus esse Terrae partem. Centrum igitur est inferior locus in universo.

Alterum argumentum est quod a partibus ipsius Terrae desumitur. Nam in cribrando tritico videmus quod glebae terrae, quae sunt in ipso tritico, ad motionem circularem cribri ad centrum ipsius cribri reducuntur; et idem evenit in partibus sabuli crassioribus, dum aliquo rotundo in vase agitantur; quo experimento multi philosophi voluerunt, Terram in medio universi stare, quia illac a motionibus coeli detruditur: quod si partibus Terrae id contingit, toti quoque Terrae id accidere dicendum est, cum in homogeneis teneat argumentum a parte ad totum, et Rothmanus in sua epistola, quae est in libro Epistolarum Astronomicarum Tychonis, pag. 185, defendens Copernicum, hoc genere argumentandi a partibus Terrae ad totam Terram utatur.

Argumenta Theologica.
Cap. 4.

Tandem, ut primam huius disputationis partem concludam, Solem non esse in medio universi, sed Terram, duo alia argumenta ex Sacris Litteris et ex doctrina theologorum desumpta, mihi ostendere videntur. Quorum alterum est ex cap. 1 Genesis, ponderando verba: Dixit Deus, Fiant luminaria in firmamento coeli. Nam, cum in textu Hebraico habeantur loco verbi *firmamento* nomen רקיע *rakia,* quod significat expansum seu extensum vel extensionem, ut probat Sanctes Pagninus in Thesauro Linguae Sanctae in radice *raka*; et huiusmodi significatio nullo modo possit convenire centro, repugnante ipsius centri natura extensioni seu, ut ita dicam, expansioni; conveniat autem aptissime coeli circumferentiae, quae quodammodo est extensa et expansa supra centrum (unde, apposita metaphora, Psalmo 103–2, dicitur, Extendens (scilicet Deus) coelum sicut pellem); dum Deus dixit, Fiant luminaria in firmamento coeli, non in centro luminare maius factum esse dicendum est, sed in ipso coeli expanso seu extenso. Confirmatur haec, argumentatio ex eo, quod verbum *Fiant,* quod Deus dixit, respicit aequaliter Solem et Lunam, cum in textu dicatur, Fiant luminaria in firmamento coeli; unde sicut Luna non est in centro, sed in coeli expanso, ita quoque Sol in hoc, et non in illo, esse debet.

Alterum argumentum est ex doctrina theologorum, tenentium ea potissimum ratione infernum, idest locum daemonum et damnatorum, esse in centro Terrae, quia, cum coelum sit locus angelorum et beatorum, oportet locum daemonum et damnatorum esse in loco remotissimo a coelo, qui est centrum Terrae. Unde bene, Psalmo 138, apponuntur infernus et coelum tanquam loca distantissima, dum dicitur: Si ascendero in coelum, tu illic es; si descendero in infernum, ades: et, Isaiae 14, dum dicitur regi Babylonis, et in eius figura diabolo: Dixisti, In coelum conscendam, etc.; veruntamen usque ad infernum detraheris, et in profundum lacum. Legatur Illustrissimus Cardinalis Bellarminus, De Christo, lib. 4°, cap. X°, et De Purgatorio, lib. 2°, cap. 6°. Cum itaque infernus sit in centro Terrae, et debeat esse locus remotissimus a coelo, Terram esse in medio universi, qui est locus a coelo remotissimus, fatendum est. Ex quibus sit impositus finis primae parti huius disputationis.

Argumenta Mathematica contra motum terrae Coperniceum.
Cap. V.

Contra motum Terrae diurnum multa obiici possunt, quorum aliqua contra Rothmanum, Coperniceae sententiae defensorem, in duabus epistolis astronomicis refert Tycho in libro Epistolarum Astronomicarum, pag. 167 et 188: videlicet de casu plumbei globi ab altissima turri perpendiculariter, non obstante praetensa aeris concomitantia, cum tamen deberet esse contrarium, quia Terra motu diurno, etiam in parallelis borealibus Germaniae, moveretur sesquicentum passus maiores in secundo minuto temporis: item de bombardis exoneratis ab oriente in occidentem et a septentrione in austrum, praesertim de exoneratis prope polos, ubi motus Terrae tardissimus est; nam, dato motu Terrae diurno, evidentissima differentia notaretur, cum tamen nulla animadvertatur.

Contra vero annuum, multo plura obiici possunt, de quibus per Tychonem ubi supra; sed ego adducam tantum quatuor rationes: quarum prima est ab ortibus et occasibus stellarum fixarum desumpta. Nam si Terra annuo motu movetur, oportet latitudines ortivas et occiduas stellarum fixarum singulis 8 aut 10 diebus sensibiliter variari; sed consequens est falsum; ergo, et antecedens. Falsitas consequentis est nota: quia latitudines praedictae non variantur notabiliter nisi in 50 aut 60 annis. Consequentia vero probatur: quia, cum Terra simul cum horizonte moveatur sub zodiaco, et sic ab austro ad septentrionem et e contra, in 8 aut 10 diebus sensibiliter, fixae vero insensibiliter propter earum tardissimum motum sub zodiaco, imo secundum Copernicum sint immobiles, necesse est ut fixae ipsae in spatio 8 aut 10 dierum notabiliter suas latitudines ortivas et occiduas varient.

Altera est ab altitudinibus polaribus locorum. Nam si Terra movetur motu annuo, oportet mutari altitudines polares locorum; sed consequens est falsum; ergo, et antecedens. Falsitas consequentis est nota. Consequentia probatur: quia, cum Terra per annuum motum feratur a septentrione in austrum et e contra, simul etiam loca in ipsa existentia sic feruntur; ista autem latio mutat omnino altitudines polares: sicut enim homini qui a meridie ad boream vel e contra iter agit, contigit altitudinem poli mutari, ita loco continget mutari altitudinem poli, si vice hominis ipse moveatur.

Tertia est ab inaequalitate dierum artificialium. Nam, etiamsi videantur omnia observationibus consentire, dato motu Terrae annuo et Solis quiete, quia horizontis rectitudo seu obliquitas eadem semper existit, cum praesupponantur horizontes simul cum Terra moveri, tamen subtiliter intuenti non ita videbitur: quia, cum per motum annuum transferatur Terra a borea in meridiem et e contra, necesse est ut zenith capitis nostri similiter transferatur, et ex consequenti ut aliquando accedat ad aequatorem et aliquando recedat; a mutatione autem zenith constat mutari rectitudinem et obliquitatem horizontis, quae inaequalitatem dierum potissimum efficit: ex quo consequitur, ut pro inaequalitate dierum signanda, non solum notandae essent differentiae motus annui, parallelos dierum artificialium efficientis, pro ut fit posito Solis motu et Terrae quiete, sed etiam illae differentiae quas efficerent mutationes obliquitatis horizontis dato motu Terrae et Solis quiete, et praecipue circa borealissimas habitationes, ubi variationes dierum sunt sensibilissimae: quod tamen non fit, cum solae primae differentiae animadvertantur, et illae observationibus satisfaciant.

Non obstat, quod horizontes simul cum Terra transferantur sine sui mutatione; quia verum esset hoc quoad motum Terrae diurnum, sed non quoad annuum: nam quoad hunc etiam quod transferantur cum Terra, tamen mutantur quoad obliquitatem et rectitudinem propter necessariam zenith mutationem, ut dictum est.

Quarta est ex Tychone in libro Epistolarum Astronomicarum, pag. 149, ubi asserit, cometas caelitus conspectos, et in Solis opposito versantes, motui Terrae annuo minime obnoxios esse, cum esse deberent, quia respectu ipsorum evanescere motum huiusmodi non est necesse, sicut in fixis syderibus, cum cometae praedicti illam maximam fixarum a Terra distantiam non habeant.

Contra denique tertium motum, Tycho in allegatis Epistolis obiicit tria: primo, quod sublato motu annuo, tertius necessario auferatur; 2°, quod fieri non potest ut axis Terrae in contrarium motui centri annuatim adeo correspondenter gyretur, ut quiescere tamen videatur; 3°, quod non potest dari in corpore unico et simplici axim et centrum duplici diversoque motu moveri; quibus si addatur etiam diurnus motus, maior efficitur difficultas.

Argumenta physica contra motum terrae.
Cap. 6m.

Plurima possent adduci argumenta physica contra Terrae motionem, quae a philosophis et a mathematicis pro Terrae quiete afferuntur, et praecipue a Tychone in allegatis Epistolis: sed ego tria tantum in medium proponam. Quorum alterum est a natura corporum gravium et levium. Nam in universum videmus, corpora gravia minus apta esse ad motum quam levia aut non gravia: quod quidem statim innotescere potest consideranti non solum simplicia corpora naturalia, sed etiam mixta, et haec non solum in ordine ad motum qui a principio intrinseco causatur, sed etiam in ordine ad motum qui fit a principio extrinseco. Rursus videmus, naturam materias ita formis accommodare, ut pro efficientia ipsarum formarum miram animadvertamus ipsarum materierum aptitudinem: et id accidit tum quia, ut dicit Philosophus, 2° Physicorum, natura agit propter finem, tum quia materiae sunt velut instrumenta formarum ad agendum. Cum itaque Terra omnium corporum nostrae cognitioni subiectorum gravissima sit, oportet dicere naturam ei tot motus nequaquam tribuisse, et praecipue diurnum, adeo velocem ut in uno minuto temporis Terra conficere debeat fere 19 milliaria, ut dicit Tycho in Epistolis Astronomicis, pag. 190.

Alterum est quod desumitur ab illa physica propositione, unicuique corpori naturali unum esse tantummodo motum naturalem; quod verum esse, inductione probari facile posset, nisi ageretur cum philosopho praestantissimo. Cum itaque Terrae motus naturalis sit ad medium, non poterit ei esse naturalis motus circa medium, et multo minus poterunt ei esse naturales tot motus, et omnes non ad medium: si igitur motus illi Copernicei non sunt Terrae naturales, quomodo fieri potest ut Terra, corpus naturale, tamdiu illis moveatur? nam naturae non est praeter naturam agere.

Tertium est ab incongruentia quadam: quia scilicet omnibus caeli partibus lucidis, videlicet planetis, motum tribuit Copernicus; Soli autem, omnium coeli partium praestantissimo et lucidissimo, motum negat, ut Terrae, opaco et crasso corpori, illum tribuat. Id enim facere non debuit, discretissima in omnibus suis operibus, natura.

Argumenta Theologica contra motum terrae.
Cap. 7m.

Argumenta theologica ex Sacris Scripturis et authoritatibus Patrum et theologorum Scholasticorum infinita possent contra Terrae motionem proponi: sed duo tantum adducam, quae firmiora mihi esse videntur. Alterum est ex Iosue, cap. X, ubi ad preces Iosue dicit Scriptura: Stetit itaque Sol in medio coeli, et non festinavit occumbere spatio unius diei; non fuit antea et postea tam longa dies, obediente Domino voci hominis. Nec responsiones, quae afferuntur, quod Scriptura loquatur secundum modum nostrum intelligendi, satisfacit: tum quia in Sacris Litteris exponendis regula est ut semper litteralis sensus salvetur, cum fieri potest, ut in nostro casu; tum quia Patres omnes unanimiter exponunt locum hunc, quod Sol, qui movebatur, re vera stetit ad preces Iosue; ab ea vero interpretatione, quae est contra unanimem Patrum consensum, abhorret Tridentina Synodus, sess. 4a, in decreto de editione et usu Sacrorum Librorum, § Praeterea. Et licet Sancta Synodus loquatur in materia morum et Fidei, tamen negari non potest, quin Sanctis illis Patribus Sacrae Scripturae interpretatio contra consensum Patrum displiceat.

Alterum est ab authoritate Ecclesiae: nam in hymno ad vesperas feriae tertiae ita canit:

> Telluris ingens Conditor
> Mundi solum qui eruens,
> Pulsis aquae molestiis,
> Terram dedisti immobilem.

Nec leve est huiusmodi argumenti genus: nam, ut videre est apud Cardinalem Bellarminum, in plerisque locis confutat multos errores hymnis, canticis et precibus Ecclesiae, quae in breviariis habentur.

Et ex his absoluta sit haec disputatio. Cui respondere aut omnino aut ex parte, videlicet saltem mathematicis argumentis et physicis, et his non omnibus sed gravioribus, tuum arbitrium esto; nam hanc scripsi non ad tentandam eruditionem et doctrinam tuam, mihi omnibusque tum in Romana Curia tum extra notissimam, sed pro investigatione veritatis, quam te semper quaerere totis viribus profiteris: et re vera sic decet mathematicum ingenium.

FINIS

PART 3
Technical Discussion of Ingoli's Essay to Galileo
and Galileo's 1624 Reply

Monsignor Francesco Ingoli's 1616 essay (part 1 of this appendix) to Galileo consists largely of "scientific" arguments against the system of Copernicus, as opposed to "religious" or "scriptural" arguments. This discussion reviews each of the scientific arguments from a technical standpoint. What does each scientific argument say? Is it scientifically valid? Was there a valid pro-Copernican counterargument? Galileo's responses from his 1624 reply to Ingoli are helpful in this discussion, and are briefly touched upon here.

In his opening preface, Ingoli describes his essay as consisting of two parts. The first contains arguments against the positions of the Earth and the Sun in the Copernican hypothesis. The second has arguments against the motions of the Earth and the immobility of the Sun in that hypothesis. He says he will use, in each part, mathematical, physical, and theological arguments. At the end of the essay, Ingoli suggests Galileo focus any response on the more weighty of the essay's scientific arguments.

Ingoli's first mathematical argument (I will number each argument for further reference—this will be M1) is based on the diurnal parallax (recall from chapter 2; see fig. 2.4) of the Moon being greater than that of the Sun. Galileo notes this to be Ingoli's own new argument,[18] as opposed to an argument Ingoli has obtained from Tycho Brahe or another astronomer. For this argument, Ingoli proceeds to discuss how astronomers note that the further removed a body is from the highest circle of heaven (that is, from the orb of the stars), the greater its diurnal parallax. In fact, diurnal parallax is determined by distance from Earth, not distance from the stars. However, Ingoli is technically correct, so long as a geocentric cosmos is assumed: in a geocentric world system, with Earth at the center of the orb of the stars, being closer to or farther from the Earth does imply being farther from or closer to the stars. Ingoli then says that in the Copernican world system, the Sun is more distant from the orb of stars (the highest circle of heaven) than is the Moon, because the Sun is in the center of the universe (the point most distant from the periphery) while the Moon is not. Applying that which works for the geocentric cosmos to a Copernican cosmos, Ingoli proceeds to declare that since, according to Copernicus, the Sun is farther from the stars than the Moon, it should exhibit greater parallax. He notes that in fact the parallax of the Moon surpasses that of the Sun by 22:1, and so therefore the Copernican hypothesis is wrong.

As Galileo notes in his 1624 reply to Ingoli, this argument simply attributes, falsely, the differences in diurnal parallax to distance from the stars, rather than to distance from the Earth.[19] A basic understanding of geometry reveals the speciousness of this argument. This argument is remarkably weak in comparison to the ones that follow it. We might speculate that Ingoli was either trying to throw Galileo an easy "first pitch," or dispensing with a very poor argument that he felt obliged to include for reasons not stated.

Ingoli's second and third mathematical arguments against the Earth's position in the Copernican system he cites as coming from two famous old astronomical texts: Johannes de Sacrobosco's thirteenth-century *Sphere* and Ptolemy's second-century *Almagest*. Both arguments are brief. The second (M2) is that Earth must be at the center of the eighth orb (the orb of the stars), because the stars appear from Earth to have the same size or magnitude regardless of their elevation above the horizon. The third (M3) is that the Earth is at the center because an Earth-bound observer always sees half of the starry heavens at any given moment. Neither would be true if Earth did not occupy the center, says Ingoli.

These arguments are valid objections. As we saw in chapter 3, if Earth is located in an orbit about the Sun, and not at the center of the universe, effects like those mentioned in these arguments—that is, differences in the stars caused by differences in position on that orbit (the phenomenon of annual parallax)—should indeed exist (see fig. 3.6). And as also we saw in chapter 3, the Copernican solution to these arguments was that the orbit of the Earth is negligibly small relative to the distances to the stars. The annual parallax effects do exist, but the stars are so distant that the effects are undetectably small. Galileo notes all this in his reply to Ingoli, saying in part that

> for Copernicus the earth is not so far from the center or so near the
> stellar sphere that the difference of a radius [of its orbit] could cause
> a noticeable change in the apparent magnitude of a star, since the dis-

tance between the earth and the fixed stars is many hundreds of times the distance between the earth and the sun.[20]

Following arguments M2 and M3, Ingoli brings up Tycho Brahe's star size objection to this Copernican solution to these arguments, referencing Tycho's *Astronomical Letters* by page number. This is discussed in detail in chapter 5.

For his fourth mathematical argument (M4) against the Earth's position in the Copernican theory, Ingoli again cites Tycho's *Astronomical Letters*. He says that there Tycho reliably proves that certain calculated technical quantities in the Copernican hypothesis are in error—in Galileo's words from his 1624 reply, "the eccentricities of Venus and Mars are different from what Copernicus assumed, and likewise the apogee of Venus is not stable, as he believed." Galileo notes that if these values are incorrect, they can be revised without affecting the basic structure of the Copernican system.[21] In other words, this fourth argument is merely a quibble over minor technical details.

Next come the physical arguments against the Earth's position in the Copernican system. The first of these (P1) is that denser bodies occupy lower positions in the universe (following the physics of Aristotle, as seen in chapter 2). Earth is denser and heavier than the sun. The lowest place in the universe is the center. Thus the Earth, rather than the Sun, would be at that center. Ingoli cites Aristotle for this argument, but he also notes that all things that have luminosity, like fire, are lightweight and low density, and move upward toward the heavens and away from the center of the Earth (which is, by definition, downward). This suggests that the luminous Sun is lightweight and of low density.

Ingoli's second physical argument (P2) against the Earth's position in the Copernican system is a behavior-of-the-parts-indicates-the-behavior-of-the-whole argument. He notes how, in wheat that is being sifted by a sieve moved with circular motion, bits of dirt in the wheat are moved toward the center of the sieve. Similarly, larger pieces in gravel that is being stirred in a round vessel are moved toward the center of the vessel. Circular motion apparently causes heavier things to collect in the center. This suggests Earth as a whole would naturally be at rest amid the circularly moving heavens.

In his reply, Galileo spends a couple thousand words to answer the roughly five hundred words of Ingoli's two arguments.[22] The problem with P2 is that a sieve's motion is not just a simple rotation, and Galileo shows how this is the case. To the much stronger P1, Galileo responds with an ingenious counterargument that he may have borrowed from Giordano Bruno: Earthly fire is brief in duration because it is a rarefied substance; since the Sun burns eternally, that shows it to be an extremely dense and solid substance.[23] A candle flame is extinguished quickly. A large piece of red-hot iron takes much longer to cool and to cease giving off light and heat. The eternally glowing Sun must be far more dense and solid than iron.

Ingoli's final arguments against the Earth's position in the Copernican system are two theological ones (T1, T2). He suggests at the end of the essay that Galileo bypass these arguments, and in his 1624 reply, Galileo does just that.

Next Ingoli proceeds to the mathematical arguments against the Copernican motions of the Earth. There are many of these, and Ingoli attributes all of them to Tycho Brahe. The first two have to do with balls dropped or balls launched from a cannon—an argument that a falling body should be deflected by the Earth's diurnal motion (M5), and the argument regarding diurnal motion and the cannon at the poles and equator (M6). In his 1624 reply, Galileo spends many pages countering these two arguments by means of the argument from common motion, which we encountered in chapter 3. Galileo describes Tycho as not understanding that a rock dropped from atop the mast of a ship retains a forward motion in common with the ship. Thus he describes Tycho as expecting the rock to drop to the rear of the mast as the ship moves forward during the time of fall, and as not understanding that this does not happen. He says Tycho likewise expects a lead ball, dropped from a tower on an Earth rotating from west to east, to fall to the west of the tower, which does not happen. Galileo also dismisses the cannon-and-the-poles argument via common motion, using a version of the shipboard cabin discussion we saw in chapter 3.[24] Recall from that chapter that the common motion argument and the shipboard cabin analogy were in fact not sufficient to answer Tycho regarding this matter.

Ingoli next proceeds to list a series of arguments from Tycho that pertain to the question of Earth's annual motion. If Earth had an annual motion about the sun, that would affect: the points of rising and setting of fixed stars (M7); the altitude of the celestial pole as seen from a given place (M8); the varying length of daylight (M9); and the trajectories of comets across the starry sky (M10). The first two of these are valid objections. They are essentially more annual parallax arguments. Effects something like what Ingoli describes would indeed be expected if Earth moved relative to the stars (fig. 3.6). Galileo objects to the way Ingoli describes them, stating, for example, that a moving Earth implies that

> there would be a change, not in the elevation of the pole, but in the elevation of some fixed star, such as, for example, the nearby Polaris, and then [an anti-Copernican could] add that, since this is not seen, one could thereby infer the stability of the earth.[25]

But Galileo goes on to say, "Copernicus already answered this by saying that, because of the immense distance of the fixed stars, such a change is imperceptible."[26] But as with arguments M2 and M3, this answer leads to the star size problem. Regarding M9, Galileo notes that in the Copernican system "the equator and its axis always keep the same inclination and direction relative to the zodiac (namely the circle of annual motion)," thus accounting for the changes in day length over a year.[27] Galileo responds to the fourth in part by criticizing Tycho, arguing that Tycho could not observe comets when in opposition to the Sun, because

> their tail always points away from the sun, [and] then it is impossible for us to see any of them. . . . Furthermore, what does Tycho know with certainty about a comet's own motion, as to be able confidently to assert that, when mixed with the earth's motion, it should produce some phenomenon different from what is observed?[28]

Ingoli completes the mathematical arguments against Earth's motion with three more arguments from Tycho Brahe. These all related to the Earth's "third motion" in the Copernican hypothesis. This third motion is that by which the Earth's axis maintains the same orientation in space, parallel to itself at all times. If one builds an orrery (fig. A.1) to illustrate the Earth's motion in the Copernican system, one needs one mechanism

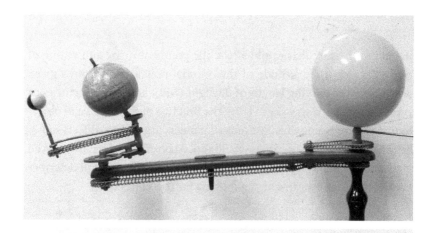

Figure A.1. A mechanical orrery that demonstrates the motion of the Earth. One mechanism moves the Earth around the Sun, another rotates the Earth, and a third (detail, bottom) keeps the axis of Earth pointed in the same direction in space. Images courtesy of Todd Timberlake, Berry College.

to cause the Earth to move around the Sun annually, another to cause the Earth to rotate diurnally, and a third to cause Earth's axis to remain pointed in the same direction in space at all times (so that, for example, the North Pole is not always tilted toward the Sun, as it might be if the axis for the diurnal rotation were to be simply fixed at an angle to the orrery's arm for annual motion). This third motion would exactly match the period of the annual motion, but be contrary to it. Today we understand the orientation of Earth's axis to be maintained through a natural rotational inertia, or tendency to not change in motion—what physicists call "conservation of angular momentum." This is the effect by which a gyroscope maintains its orientation in space. But at the time, many considered the orientation of Earth's axis to be maintained through an actual third motion.

The first of these third motion arguments is that no third motion is needed if there is no annual motion (M11). The second is that such a third motion could not be precisely contrary to the annual motion (M12). The third is that a single body cannot have so many simultaneous motions (M13).

Galileo responds to these arguments by citing the example of a wooden ball floating in a bowl of water. A person holding the bowl can turn around, and to that person the ball appears to turn in the water. However, the ball actually maintains its orientation in space and does not truly turn with its own motion.[29] In a sense, Galileo recognizes that bodies have a rotational inertia. Thus he dismisses the whole idea of a third motion.

For the physical arguments against Earth's motion in the Copernican system, Ingoli begins by again referencing Tycho Brahe. The first of these arguments (P3) is that heavy bodies are unsuited to motion, and so it is not proper for the Earth to have all the motions it has in the Copernican system. As we saw in chapter 3, that the Earth was too heavy to be suitable for motion was an important point for Tycho. Galileo answers this argument by, among other things, denying that heavy bodies are unsuitable to motion. He notes that, when fired by a cannon, a lead ball travels farther than a wooden ball, which in turn travels farther than a wad of straw or fiber. He remarks, "I see bowl makers and tin-plate turners adding very heavy wooden wheels to their machines in order to make them retain longer the impetuses they acquire"[30] (see chapter 2 regarding impetus). Thus Galileo, much like Buridan, illustrates that heavy bodies can actually retain motion.

The next physical argument (P4) that Ingoli lists is based on the proposition that says that only one natural motion belongs to a natural body. Since the natural motion of earth (in Aristotle's system of elements) is toward the center of the universe, the Earth as a whole cannot also have a natural motion around the Sun. Galileo writes in his reply a lengthy discussion about straight and curved motion, ending with "So I conclude that if the earth has a natural inclination to motion, this can only be toward circular motion."[31]

The third and last of these physical arguments (P5) is that Copernicus lacks consistency in terms of which sorts of bodies move—the bright planets move, the dark Earth moves, but the brilliant Sun is at rest. Consistency suggests that bright heavenly bodies all move, while dark bodies like Earth do not. Galileo's 1624 reply was to point out that indeed all these bodies except the Sun are devoid of light (as inferred from observing the phases of Venus and the Moon, for example), while the Sun is the source of light, and that indeed Copernicus is very consistent here. "Therefore we can very resolutely assert that the earth's conformity to the other six planets [Mercury, Venus, the Moon, Mars, Jupiter, and Saturn] is very great, and that on the contrary the discrepancy between the sun and these bodies is equally great."[32]

Ingoli's final arguments against the Earth's motion in the Copernican system are two more theological ones (T3 and T4). Again, in the closing paragraph of his essay Ingoli suggests Galileo focus on the weightier mathematical and physical arguments, rather than the theological arguments. In his 1624 reply, Galileo bypasses these arguments altogether.

Thus Ingoli's essay includes thirteen mathematical arguments, five physical arguments, and four theological arguments. In addition, the essay includes a discussion of Tycho's star size objection to the Copernican hypothesis, which Ingoli does not list as an argument per se. The "scientific" arguments are of varying quality. Some, such as M1 (about the diurnal parallax of the Sun and Moon) and P5 (about consistency among bright and dark bodies) were quite weak, and Galileo could easily dismiss them on the basis of simple observations and basic geometry. Others, such as P1 (about the positions of lightweight and heavy bodies, and the lightweight nature of the Sun) and P3 (about the unsuitability of heavy bodies for motion) were solid arguments, if one accepted the standard Aristotelian physics of the day and certain casual observations about nature (those things that give off light, such as flames, tend to be lightweight and rise to the heavens; those things that are heavy, such as loaded wagons, are difficult to move). However, Galileo could answer them by appeal to a new, albeit undeveloped, physics and other observations about nature (iron heated by a blacksmith gives off light, and the larger the amount of iron the longer it takes to cool and cease glowing; a heavy projectile is more suitable to be thrown than a very lightweight one). But still others among the scientific arguments were strong enough—weighty enough, to use Ingoli's term—that Galileo's answers either did not sufficiently address the argument (M6, the cannon, poles, and equator argument) or they introduced the star size objection of Tycho (M2, M3, M7, M8, all arguments related to parallax).

Appendix B

A Rendition into English and a Technical Discussion of
Giovanni Battista Riccioli's Reports Regarding His Experiments
with Falling Bodies

What follows in part 1 of this appendix is an effort to render Riccioli's Latin into a form relatively accessible to the modern reader. As in appendix A, while we (Christina Graney provided invaluable assistance in translating this work) strove to produce a faithful rendering of the work, there were places where we felt the need to break the essay into additional paragraphs, or to paraphrase somewhat. Also, we have frequently divided lengthy Latin sentences into multiple shorter English sentences. Our paragraph breaks are indicated by indentations; original paragraphs are indicated by line spaces. We have generally retained the original capitalization and italics, which do not follow modern practice. For the reader who wishes to see the original Latin, it is included in part 2 of this appendix. The Latin is taken from Riccioli 1651, 2:384–89. Lastly, part 3 of this appendix consists of a technical discussion of Riccioli's reports.

PART 1
English Rendition of Riccioli's Reports Regarding
His Experiments with Falling Bodies

II. The Group of Experiments about Unequal motion
of Heavy Bodies descending faster and faster in the Air,
by which they come nearer to the end to which they tend

VI. The *first* Experiment is taken from sound. Let a ball of wood or bone or metal fall from a height of 10 feet into an underlying bowl, and attend to the ringing arising from the percussion. Then let that same ball fall from a height of 20 feet, and indeed you will perceive a far greater and more extensive sound poured out. Afterward lift up the bowl to a height of 10 feet and into that let drop the same ball from an altitude of 10 feet above the bowl. Indeed you will perceive a ringing like at first. Therefore that ball in the second fall has acquired a greater impetus[1] because of the drop from a greater height, than from the smaller heights in the first and third falls. And in the second fall the ball has gained more impetus in the second half of the journey down than in the first half, by as much as its downward velocity will have increased in the latter 10 feet of the fall, than in the former 10 feet.

In fact it is manifest by continual experiments that in a moving body greater impetus accompanies swifter motion. A household experiment with this phenomenon is to pour out water into a ladle from a vessel about two or three fingers' breadth from the ladle. You will perceive no noise. Elevate the vessel to two or three feet, and you will perceive noise from the falling water. Hence *Cicero* in Somnio Scipionis reports that people near the cataracts of the Nile have been deafened because of the crashing of the water falling from the precipitous height.

The *second* Experiment is taken from the impact perceptible by the sense of touch. Place your hand below a ball while someone lets it fall from an altitude of 10 feet. Indeed you will experience the lightest impact. But if the same ball is let fall from an altitude of 50 feet or greater, your hand will perceive some pain from the impact: therefore a greater impetus is derived from a higher fall. Poor Aeschylus[2] sensed this greater impetus that I have described, from the turtle that the Eagle dropped onto his head; and the stupid bird herself felt no doubt concerning what would happen. Elpenor[3] sensed it in falling from the tower. So Ovid writes in the third book of *Tristia*

> *Who falls on level ground—though this scarce happens—*
> *so falls that he can rise from the ground he has touched,*
> *but poor Elpenor who fell from the high roof*
> *met his King a crippled shade.[4]*

And from this source we know that adage and well-known warning of the Poet

> *. . .The higher they are raised,*
> *The harder they will fall . . .*

Finally, is it not true that they who run down a slope receive so great an impetus that, however much they may wish, they cannot stop their forward movement at the bottom, even though they could easily stop it at the beginning?

VII. The *Third* Experiment is taken from the greater impact of falling from a height, but estimable by the eyes: namely, a clay ball released from a small altitude is not itself broken, neither can it break the shell of an egg or the hull of a nut placed perpendicularly under it, nor can it elevate a weight placed in a wooden two-pan balance [when it strikes the other pan], nor penetrate a palm's-depth of water; if it is released from a higher place, it does all those things: namely, it breaks, and it is broken, and it elevates that weight. Thus a wooden ball, or playing ball, falling from low altitude into a cistern or a large vessel full of water, is immersed a few finger's breadth underneath the water; but if that ball may plummet from a very high place, it penetrates to many feet below the water and finally all the way to the bottom. And by other innumerable experiments of this sort it is made evident that a heavy body falling from a higher place always naturally accrues greater and greater impetus at the end of its motion.

The *Fourth* is taken from bouncing, and by the rebound in the height of a playing ball. Indeed we arranged for a very hard leather ball of this sort, no greater in size than the yolk of an egg, to be released in order that it might fall to the earth at an acute angle from an altitude of 37 feet, down to the flat stone pavement. It rebounded all the way to 7½ feet. When it was released from an altitude of 73 feet it rebounded to 11¼ feet. A larger leather ball released from an altitude of 37 feet and hitting the pavement by a more obtuse angle rebounded to nearly 6 feet; released from an altitude of 73 feet it rebounded to 7½ feet.

But I see myself as playing, as long as I am not progressing toward more noble experiments, and toward clearly demonstrating not only the nonconstant motion of heavy bodies, but also the true growth of their velocity, which increases by uniform differences as the motion progresses.

VIII. Then the *Fifth* Experiment often taken up by us has been the measuring of the space that any heavy body traverses in natural descent [free fall] during equal time intervals. This was tested at Ferrara by Fr. [Niccolò] Cabeo in 1634, but only from the tower of our church there (less than 100 feet in height) and using an uncalibrated Pendulum.

But in 1640 in Bologna, I calibrated Pendulums of various lengths using the transit of Fixed Stars through the middle of the heavens. For this [free fall] experiment I have selected the smallest one, whose length measured to the center of the little bob is one and fifteen hundredths of the twelfth part of an old Roman foot, and a single stroke of which [that is, the half period] equals one sixth of a second, as I have shown and set forth in book 2, chapter 21. As a single Second exactly equals six such strokes, then one single stroke is nearly equal to that time with respect to which the notes of semichromatic music are usually marked, if the Choirmaster directs the voices by the usual measure.

The oscillations or strokes of so short a Pendulum are very fast and frequent, and yet I would accept neither a single counting error nor any confusion or fallacious numbering on account of the eye. Thus our customary method was to count from one to ten using the concise words of the common Italian of Bologna (*Vn, du, tri, quatr, cinq, sei, sett, ott, nov, dies*), repeating the count from one, and noting each decade of pendulum strokes by raising fingers from a clenched hand. If you set this to semichromatic music as I discussed above, and follow the regular musical beat, you will mark time as nearly as possible to the time marked by a single stroke of our Pendulum. We had trained others in this method, especially Frs. Francesco Maria Grimaldi and Giorgio Cassiani, whom I have greatly employed in the experiment I shall now explain.

Grimaldi, Cassiani, and I used two Pendulums; Grimaldi and Cassiani stood together in the summit of the Asinelli Tower [in Bologna], I on the pavement of the underlying base or parapet of the tower; each noted separately on a leaf of paper the number of pendulum strokes that passed while a heavy body was descending from the summit to the pavement. In repeated experiments, the difference between us never reached one whole little stroke. I know that few will find that credible, yet truly I testify it to have been thus, and the aforementioned Jesuit Fathers will attest to this. That is all concerning the Pendulum and the measure of time.

After this we prepared a very great basket full of clay balls, each of which weighed eight ounces. For the shorter distance measures at least, we used the windows of our College. But for the higher ones, we used windows or openings of different towers: the tower of St. Francis, which is 150 Roman feet high; and St. James, which is 189 feet; and St. Petronius of Bologna, which is 200 feet; and St. Peter, which is 208 feet; but especially and more frequently the tower of Asinelli, which is 312 feet high altogether, and 280 feet from the summit to the base or parapet. The Asinelli is as commodious as possible to this sort of experiment, just as if it might have been constructed for this purpose. It is a delight to the eye.

IX. Shown in [figure B.1] is the rather thick trunk of the Tower IBCD over an almost cubic base VYZX that is much broader than the trunk of the tower. From this base the parapet YZH stands out, fenced in by wide stone railings. At least six men may safely walk abreast around the tower, between it and the railings. On top, the crown BC stands out, fenced in by peaked stone railings. Thus any man of ordinary stature may be able to look over safely from the railings at G, Φ, and O, as well as from the windows. From there a plumb line can be dropped all the way to the pavement of the parapet ID to measure the height GI. Fr. Grimaldi and I have done this more than once, obtaining the value I stated earlier of 280 old Roman feet. We measured the rest as well.

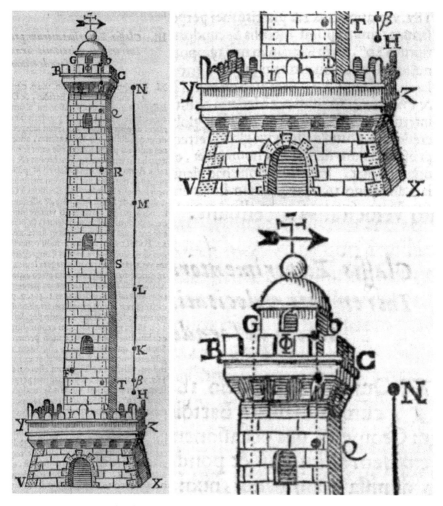

Figure B.1.   Riccioli's diagram of the Asinelli tower in Bologna (Riccioli 1651, 2:385). Image credit: ETH-Bibliothek Zürich, Alte und Seltene Drucke.

Thus this sort of Tower has great advantages for this sort of work. For instance, balls released from openings G, Φ, and O fall perpendicularly to the pavement ID, neither impinging on the foot of the tower, nor falling outside the railings YZ. The balls do not fall out into the street by the base of the tower. Balls can be released from the crown often and frequently without danger to anyone. In addition, the tower has iron belts around F and T, with iron clasps, constraining opposite walls. We have used these as reference points for measurements as well. The line NH indicates heights that we have used, including those at other towers.

X. So in May of 1640, and at other times afterward, we determined the height Hβ, that from which an eight ounce clay ball, when released, will strike the pavement at precisely five exact strokes of the pendulum described above (that is, 5/6 of a second of time). Through oft-repeated experiments we have discovered this to be 10 Roman feet. Then we determined the height necessary for a ball of the same type and weight to descend in twice as much time, or ten strokes. We discovered this to be 40 feet, which interval KH marks. Ascending further, we determined the appropriate height for thrice as much time, or exactly 15 strokes, which we discovered to be 90 feet, LH. We discovered that for a time of 20 strokes the height MH to be 160 feet, and for 25 strokes the height NH to be 250 feet. Finally, we could not ascend sufficiently high for a ball to require 30 strokes for its descent. So instead we repeatedly released a ball from the crown of the Asinelli tower at G to strike the pavement at I, which is a distance of 280 feet. With me at I and Frs. Grimaldi and Cassiani at G, we consistently counted 26 strokes, as we discovered by comparison of our written notes.

Now let us imagine the intervals marked on the line NH, translated to intervals on the line OT. The distance the ball has travelled at the end of the first five strokes, OC, is 10 feet, and equals βH; the distance the ball has travelled at the end of the second five strokes, OQ, is 40 feet, and equals KH; at the end of the third five, OR, equal to LH, is 90 feet; at the end of the fourth set OS, equal to MH, 160 feet; at the end of the fifth set or 25 strokes the total OT equals the whole NH, 250 feet. Based on the motion prior to T you have some indication of what occurs as the ball continues into the pavement.

Therefore the aforementioned ball descends faster and faster the farther it recedes from O and the nearer it approaches to D. In terms of equal measures of time, during the first measure it traverses OC, 10 feet; during the second it traverses CQ, 30 feet; third, QR, 50 feet; fourth, RS, 70 feet; and fifth, ST, 90 feet. These numbers sum to 250 feet. Any such conspicuous and noted growth is deserving confirmation by further experiment. Indeed we examine this alone: whether Heavy bodies falling naturally through the air in a straight line perpendicular to the horizontal, descend at a speed that is uniform, or increasing or decreasing by uniform differences.

III. The Group of Experiments about the Proportion of the Growth
of the Velocity of Heavy Objects descending through the Air

XI. I did not understand or recognize the proportion of the growth of the velocity of Heavy bodies related by Galileo in the Second Day of his Dialogue of the system of the World, and asserted by him to be following odd numbers begun from unity. This is true even though I might have discovered it myself, beginning in at least the Year 1629 during an occasion when I was with Fr. Daniel Bartolo, and Dr. Alphonso Iseo the eminent Geometer (examining two pendulums of the same height and weight simultaneously released from the same terminus, and whether they might always advance by like pace through any number of swings; noting how all oscillations of the same pendulum are mutually equal in time, or synchronous), and again later in 1634 with Fr. Cabeo at Ferrara. Indeed, at that time, according to my ruder experiments, I suspected it to be continually triple, as in these numbers: 1, 3, 9, 27.

Yet still later the opportunity was granted to me of reading Galileo's dialogues, which the Holy Congregation of the Index had prohibited. I found in the dialogues on page 217 of the Italian or 163 of the Latin[5] the aforementioned growth, discovered by experiment, to be following simple odd numbers from unity, as in 1, 3, 5, 7, 9, 11, etc. Still, I was suspecting something fallacious to lurk in the experiments of Galileo, because in the same dialogue, following page 219 of the Italian, 164 of the Latin,[6] he asserts an iron ball of 100 Roman pounds released from an altitude of 100 cubits reaches the ground in 5 Seconds time. Yet the fact was that my clay ball of 8 ounces was descending from a much greater altitude, namely from GI (280 feet, which is 187 cubits) in precisely 26 strokes of my pendulum: 4 and 1/3 Seconds time. I was certain that no perceptible error existed in my counting of time, and certain that the error of Galileo resulted from times not well calibrated against transits of the Fixed stars—error that was then transferred to the intervals traversed in the descent of that ball. Furthermore, I was scarcely believing that Galileo had been able to use an iron ball of such great weight, especially when he did not even name the tower from which he might have arranged for such a ball to be released.

And so, full of this suspicion, I began exacting measure of this growth in the Year 1640, as I have said. I hoped to contrive my own idea about this that was nearer to the truth; but rather I have in fact discovered to be true that which Galileo asserted. And indeed as I set forth in the preceding experiment (paragraph number X), I have acknowledged the growth to follow the proportion of 10, 30, 50, 70, 90 feet, which expressed in smallest numbers is just 1, 3, 5, 7, 9.

But not yet completely acquiescing to that, I examined with Father Grimaldi the height required in order that the eight ounce clay ball when released might reach the pavement in 6 strokes of the pendulum, or one Second; we obtained the height βH to be 15 feet. For two Seconds, or 12 precise strokes, the height KH, was 60 feet. For three Seconds or 18 strokes the height LH was 135 feet. For 24 strokes or four Seconds the height MH was 240 feet. The same ball again and often transited the 280 feet at the Tower of Asinelli in 26 strokes or 4 and 1/3 Seconds.

I have discovered the preceding number of strokes to exactly correspond to the preceding distance intervals, although in the greater distances one or another foot less or more does not introduce a difference of one whole stroke. Hence this experiment shows the proportion for distances traversed in equal times to have been 15, 45, 75, 105 feet, which follows 1, 3, 5, 7 (for as 1 is to 3, so is 15 to 45, etc.). The results of these two experiments (and a third which I have not discussed here, both for the sake of brevity and because fraction numbers are involved) are found in [table B.1].

Therefore Fr. Grimaldi and I went to talk to the distinguished Professor of Mathematics at Bologna University, Fr. Bonaventure Cavalieri, who was at one time a protégée of Galileo. I told him about the agreement of my experiments with the experiments of Galileo, at least as far as this proportion. Fr. Cavalieri was confined by arthritis and gout to a bed, or to a little chair; he was not able to take part in the experiments. However it was incredible how greatly he was exhilarated because of our testimony.

XII. Now those not ignorant of Geometry recognize the distances traversed by naturally descending heavy bodies of this sort increase as the squares of the elapsed times of descent. Galileo himself notes this in day 2 of the Dialogue page 217 or 163.[7]

For instance, in the first experiment, the strokes followed the sequence 5, 10, 15, 20, 25, the square numbers of which (that is, the products of the same number with itself) are 25, 100, 225, 400, 625; while the distances traversed followed the sequence 10, 40, 90, 160, 250. Now as 25

Order of Experiments	Strokes of a Pendulum of length 1 & 15/100 inches	Time corresponding to the strokes (in seconds)	Square of the number of Strokes	Distance traversed by an 8-ounce clay ball at the end of the times (in Roman feet)	Distance traversed during each equal interval of time (in Roman feet)	Proportion of the growth of the velocity of heavy bodies in air (smallest numbers)
1	5	50/60	25	10	10	1
	10	1 & 40/60	100	40	30	3
	15	2 & 30/60	225	90	50	5
	20	3 & 20/60	400	160	70	7
	25	4 & 10/60	625	250	90	9
2	6	1	36	15	15	1
	12	2	144	60	45	3
	18	3	324	135	75	5
	24	4	576	240	105	7
	26	4 & 20/60	676	280	40	8 & 1/6
3	6 & 1/2	1 & 5/60	42.25	18	18	1
	13	2 & 10/60	169	72	54	3
	19 & 1/2	3 & 15/60	380.25	162	90	5
	26	4 & 20/60	676	280	118	6 & 1/18

Table B.1. New *Almagest* data table showing times and distances for eight-ounce clay balls dropped from varying heights. The highlighted values are fully independent measurements (see part 3 in this appendix).

is to 100, so 10 is to 40; as 100 is to 225, so 40 is to 90; as 225 is to 400, so 90 is to 160; finally as 400 is to 625, so 160 is to 250. Thus in experiment 2, the order of Strokes was 6, 12, 18, 24, 26 the squares of which are 36, 144, 324, 576, 676. Truly the distances traversed were in sequence 15, 60, 135, 240, 280. But as 36 is to 144, so 15 is to 60; as 144 is to 324, so 60 is to 135; as 324 is to 576, so 135 is to 240; finally as 576 is to 676, so 240 is to 280. Therefore the preceding distances have increased in proportion to the squares of the strokes (that is, times). Reducing those times to least numbers, so that the first time is 1 unit, the second 2, the third 3, the fourth 4, and the fifth 5, the squares progress as 1, 4, 9, 16, 25.

In the third experiment (which you have available in the table seen here) the number of feet [on the last entry] ought to be exactly 288 in order that the preceding proportion might be preserved. However, the greatest height we were able to test was 280 feet.

It is truly pleasing to collect all that we have thus far said about so beautiful a proportion, and the basis of it, into one table that provides the Reader with a short glancing synopsis. Yet I add that at one time I was hoping we would discover this same proportion in the weight elevated by a falling ball [striking one side of a balance]—so that if a ball falling from a height of one foot raised a weight an inch, it would raise the weight doubly much from the height of 4 feet, or triply much from the height of 9 feet. It did not. See the upcoming group XII of the Experiments.

IV. The Group of Experiments about the Unequal Descent of Two Heavy Bodies of unlike weight through the Air from the same altitude

XIII. This experiment is very misleading unless it is completed with great diligence and circumspection. All present must be able to reach the same conclusions concerning which motions will be faster or slower.

The best arrangement, which would remove all doubts, would be if two balls of the same material and same appearance but different weights would be released from the same height at the same time. Imagine that we could take two clay or metal balls—one a palm's width in diameter and weighing one Roman pound, the other two palms and eight pounds—and rarefy the smaller one while retaining its solidity (without adding any foreign material to it) so that it became the size of the larger. Then if in fact the heavier was found to descend more quickly, the quicker descent could be attributed only to the strength of greater heaviness, and not to difference of appearance or bulk, nor to difference of shape (where a larger ball would have to break through the broader expanse of air below it). In short, the best arrangement is one of parity in the case of all things except weight.

It is truly not expedient to release two cylinders or two prisms made from the same material, of equal width, but of unequal height or weight. They fluctuate around their centers of gravity in the air, and they yield uncertain cause of swifter or tardier descent.

The best arrangement is not possible, and so Fr. Grimaldi and I prepared what is nearest to it: twelve clay balls of solid clay, of which the weight of each was 20 ounces; and twelve others from clay, equal in bulk to the 20 ounce balls, but hollow on the inside, each weighing 10 ounces. Then we acquired many other diverse balls, differing in bulk, or in weight. We then prepared different comparisons.

In May 1640, August 1645, October 1648, and most recently in 1650 we released different pairs from the crown of the tower of Asinelli, before many witnesses convoked from our Society.[8] Although they have each not always been to the same experiment, these include: *Frs. Stephanus Ghisonus, Camillus Rodengus, Iacobus Maria Pallavicimus, Franciscus Maria Grimaldus, Vincentius Maria Grimaldus, Franciscus Zenus,* and *Paulus Casatus* with his students *Franciscus Adurnus* and *Octavius Rubeus.* Each is distinguished by character, judgment, and religious integrity. And indeed among these, three or four Masters of Philosophy or Theology were present, who with Galileo, Cabeo, or Arriaga, had judged that any two heavy bodies, released simultaneously from the same altitude, however great, descend to the ground in the same natural moment of time.

But they promptly gave up this opinion, since a lighter (10 ounce) clay ball—released from G [see fig. B.1] at the same moment as a second clay ball of the same bulk but of 20 ounces was released from O—was seen to be at F, distant from the pavement I by at least 15 feet, at the same moment at which the heavier ball struck the same pavement at D and burst into myriad fragments. Although before the arrival of each to the pavement the lighter ball (which by agreement Fr. Grimaldi always released from the left hand) would appear greatly separated from the heavier one a little below the middle of the tower. In terms of time, counted by the help of the pendulum, the heavier ball descended by exactly 23 strokes, or 3 and 1/2 seconds,[9] the lighter by 26 strokes, or 4 and 1/3 seconds. This experiment was repeated twelve times and always had a similar outcome.

Since the distinction between the two balls was more easily observed in the difference of distance than in the difference of time, the decision was to attend to the residual distance interval FI of the more slowly descending ball, than to the time interval, as seen in [table B.2] (in which the earlier experiments [listed in the table] consider two balls of the same bulk, while the later experiments consider balls of different bulk).

Conclusions from [Table B.2]

XIV. Here I suppose one body to have greater specific Heaviness than another if, bulk being equal, it is greater in weight, as lead is said to be heavier than wax, because if two palm-wide balls may be weighed, one lead, the other wax, the lead will weigh either more pounds or ounces or drachmas etc. than the wax. By contrast, a body has greater individual Heaviness, whether it is of the same specific heaviness or not, if it is of greater weight absolutely, just as a wax ball of a hundred pounds is said to be heavier than a lead ball of one ounce. Hence five useful combinations arise:[10]

1. Those where both objects are equally heavy specifically and individually—for example, two lead spheres, of one pound each. These are naturally of the same bulk. If one were rarified supernaturally or preternaturally, it could reach a greater bulk.
2. Those where both objects are equally heavy specifically, but not individually—for example two lead balls, where one weighs one, the other two pounds. If these both be solid, then that which is the heavier individually—the two pound ball—is also the larger; by supernatural or preternatural compaction it might be made to be equal to or lesser than the other. And the lighter ball might equal the heavier one in bulk if made hollow on the inside, or full of air.
3. Those where both objects are equally heavy individually, but not specifically—for example, a lead ball and a wax ball, each of which is one pound. Naturally that which is the lighter specifically (the wax) is

Table B.2. *New Almagest* data tables for pairs of balls dropped from the same height, showing which one reached the ground first.

		Balls released at the same moment from an altitude of 280 Feet			
Order of the Experiments	Balls of the same size	Weight of the balls Ounces	Drachmas [1/8 ounces]	Ball which was faster	Distance of the slower from the pavement when the faster struck (feet)
1	Clay hollow	10	0		
	Clay solid	20	0	Heavier	15
2	Clay hollow	20	0		
	Clay solid	45	0	Heavier	16
3	Clay	9	0		
	Wood	2	4	Heavier	20
4	Clay	20	0		
	Wax	15	0	Heavier	12
5	Wood	4	6		
	Wax	6	7	Heavier	15
6	Wax	1	5		
	Iron	11	7	Heavier	30
7	Clay	5	0		
	Clay	4	0	Heavier	5
8	Clay	21	4		
	Clay	11	4	Heavier	12
9	Clay	27	3		
	Clay	14	1	Heavier	15
10	Clay	18	7		
	Wax	1	0	Heavier	35
11	Clay	2	0		
	Lead	2	4	Heavier	25
12	Lead	2	4		
	Wood	2	4	Lead	40
13	Clay	7	0		
	Clay	62	0	Heavier	10
14	Clay	10	4		
	Clay	23	0	Heavier	13
15	Clay	53	0		
	Clay	6	4	Heavier	8
16	Clay	53	0		
	Clay	7	1	Heavier	9
17	Walnut Wood	2	1		
	Beech Wood	4	7	Walnut	2
18	Clay	11	0		
	Lead	1	7	Lead	1
19	Clay	33	0		
	Lead	1	0	Clay	2
20	Clay	38	0		
	Lead	1	0	Clay	3
21	Lead	1	0		
	Lead	0	4	Heavier	5

Experiment 1 was repeated 12 times, experiment 2 was repeated twice.

the larger; if it could be compacted supernaturally or preternaturally it could be made to be no bigger than the other.

4. Those where one object is Heavier both specifically and individually— for example a 10 pound lead ball compared to a one pound wax ball. The heavier object can be of equal bulk, or lesser, or greater.

5. Those where one object is heavier specifically, but lighter individually—for example, a little lead ball of one ounce and a wax ball of one hundred pounds. In such a case the first is naturally smaller than the second. Here the lead ball is heavier specifically, while the wax ball is heavier individually.

Granted these, the conclusions below follow from [table B.2].

First: Two spheres, equally heavy both specifically and individually, naturally descend equally rapidly through the same medium (if released simultaneously from the same altitude). They are of equal bulk, so all things are like. Yet suppose one of those might be made (supernaturally or preternaturally) either greater by rarefaction, or lesser by compaction: the smaller sphere would descend faster; the contact angle, by which it might hit the plane of the underlying air, would be more acute.

Second: Of two spheres equally heavy specifically but not individually, the one that is heavier individually naturally descends more quickly through the air. This is true whether they are of equal bulk (because one of them is hollow within), as in experiment 1 (repeated twelve times) and experiment 2; or of differing bulk, as in experiments 7, 8, 9, 13, 14, 15, 16, and 21 [see table]. Some of these I believe may indicate unknown proportional relationships between speed and the individual heaviness, the size, and the lack of roundness of a ball. However, it is better to report experiments faithfully rather than touch them up by selecting out certain results.

Third: Of two spheres equally heavy individually, but not specifically, the one that is heavier specifically naturally descends more quickly through the air, as seen in experiment 12. The reason for this is that the one that is heavier specifically is naturally of smaller bulk than the other. Thus the one that is heavier specifically is also sharper of shape or angle of contact (owing to the smallness of the sphere). This is apart from the fact that, in our case, the wood sphere is more porous than the lead, and less suitable to gravitating, on account of the levity[11] of the air or vapors hidden within the pores.

Fourth: Of two spheres, one of which is heavier not only specifically but also individually, the heavier one naturally descends more swiftly through the air, whether it is larger than the other, as in experiment 10, or equal to the other, as in experiments 3, 4, 5, and 6, or indeed smaller, as in experiment 11.

Fifth: Of two spheres, of which one is heavier specifically but not individually, the one that is heavier specifically may descend more quickly, equally quickly, or less quickly through the air. We see an example of the first case in experiments 12, 17, and 18; of the third case in experiments 19 and 20. Moreover we have sufficient argument for the second and third by comparing experiment 8 with 18.

Indeed, in both, the Clay ball was 11 ounces (neglecting drachmas). But in experiment 8 the clay of 11 ounces released with a clay of 21 ounces was slower, being distant 12 feet from the pavement when the 21 ounce clay struck; whereas in experiment 18 the clay of 11 ounces released with the lead of almost 2 ounces was not slower, except by just an interval of one foot. Therefore if the clay of 21 ounces might have been released with the lead of almost 2 ounces, the lead might have trailed it by an interval of 11 feet (or at least by more than one foot). The 21 ounce clay achieved the faster descent.

Therefore with increased individual weight the speed of descent increases to faster from slower. Certainly the individual weight of the 11 ounce clay could be increased a little so as to descend equally swiftly with the lead. *Anyhow* I have said *more than enough.*

Experiments 19 and 20 indeed startled me. In these a little lead ball of one ounce acquired more (or at least nearly as much) velocity on account of smallness than the heavier clay acquired on account of far greater weight. The more perfect roundness of the lead than of the clay contributed to this.

Sixth: Of two balls simultaneously released from the same altitude through the same air, that ball that is lighter in every respect at no time naturally descends more quickly or equally quickly. Indeed, the swifter is either heavier specifically and individually, or heavier individually, or heavier specifically. Hence, said simply, those that are heavier naturally descend more quickly, at least in our air. However, it can happen that two balls are of different weight, and the lack of individual heaviness in one may be compensated for by its sharpness of small bulk, and they descend equally swiftly.

PART 2
Riccioli's Reports Regarding His Experiments
with Falling Bodies

*II. Classis Experimentorum Pro Inaequali motu Grauium in
Aere velocius ac velocius descendentium, quo plus accedunt
ad terminum, ad quem tendant.*

VI. *Primum* Experimentum sumitur ex sonitu, dimitte enim ex altitudine pedum 10. globum ligneum, vel osseum, vel metallicum in subiectam peluim, & aduerte ad tinnitum ex percussione ortum. Deinde illum
ipsum globum dimitte ex altitudine pedum 20. seu duarum perticarum,
senties enim longe maiorem ac latius diffusum sonitum: Posteà eleua
peluim ad altitudinem pedum 10. & in illam dimitte eumdem globum ex
altitudine pedum 10. senties enim tinnitum primo similem. Ergo globus
ille in secundo casu maiorem impetum acquisiuit ex lapsu ab altitudine
maiori, quam ex minori in primo & tertio casu; & in secunda medietate
itineris deorsum, plus impetus acquisiuit, quam in prima medietate,
atq. adeo inaequali velocitate deorsum motus est, & ita inaequali, vt velocius descenderit in secundo casu per 10. posteriores, quam per 10. priores
pedes. Siquidem perpetuis experimentis manifestum est ad maiorem impetum sequi velociorem motum mobilis. Quod si domesticum magis experimentum vis, insunde cyato aquam ex vase ad duos tresue digitos dis-
tante a cyato, nullum strepitum senties, eleua vas ad duos tresue pedes, &
strepitum senties ex aqua cadente; Hinc qui ad catadupas, seu cataractas
Nili habitabant obsurduisse dicuntur a *Cicerone* in Somnio Scip. ob frago-
rem aquarum praecipiti lapsu ex alto ruentium.

Secundum sumitur ex percussione tactu ipso perceptibili. Suppone
manum pilae lusoriae dum ab altero demittitur ex altitudine pedum 10.
experiere enim leuissimam percussionem, at si ea demittatur ex altitudine 50. pedum aut maiori, senties non sine aliquo dolore percuti manum
tuam: maiorem ergo impetum concepit ex altiore lapsu. Quem maiorem
impetum sensit, vt dicebam, miser Aeschylus ex testudine, ab Aquila de
sublimi deiecta in ipsius caput; & praesensit bruta ipsa volucris, nihil de
successu futuro dubitans. Sensit Elpenor ex turri prolapsus, vnde Ouidi-
ana illa comparatio lib. 1. de Tristibus.

Qui cadit in plano (vix hoc tamen euenit ipsum)
Sic cadit vt tacta surgere possit humo.
At miser Elpenor tecto delapsus ab alto,
Occurrit Regi flebilis vmbra suo.

Et hinc illud a dagion, ac notissimum Poetae monitum

> *. . . .Tolluntur in altum*
> *Vt lapsu grauiore ruant. . . .*

Demum nonne ij, qui currunt deorsum ex aliquo cliuo tantum impetum concipiunt, vt quamuis velint, nequeant inhibere postea cursum, quem ab initio facillime inhibere poterant?

VII. *Tertium* Experimentum sumitur ex percussione maiore cadentium ex alto, sed oculis aestimabili: globus enim argillaceus, qui ex parua altitudine dimissus non frangitur ipse, aut non valet frangere testam oui aut corticem nucis subiectam ipsi perpendiculariter, aut eleuare pondus in bilanci lignea collocatum, aut peruadere palmum profunditatis aqueae; si ex altiore loco dimittatur, omnia illa tandem praestat: frangit enim, & frangitur, & eleuat pondus illud. Sic globus ligneus, aut pila lusoria ex humili altitudine in cisternam, vel magnum vas aqua plenum, decidens paucos digitos infra aquam mergitur at si ex sublimi valde loco cadat, ad multos pedes infra aquam & aliquando ad fundum vsque peruadit. Et alijs innumerabilibus experimentis huiusmodi euidens fit, maiorem semper ac maiorem impetum in fine motus acquisiuisse corpus graue naturaliter cadens ex altiore loco.

Quartum sumitur ex repercussione, et resultu pilae lusoriae in altum: Etenim pilam huiusmodi coriaceam praeduram, et magnitudine non maiore vitello oui, vt acutiori angulo terrae incideret iussimus dimitti ex altitudine pedum 37. in pauimentum lapide stratum, eaque resultauit vsque ad pedes 7½. at quando dimissa fuit ex altitudine pedum 73. resultauit ad pedes 11¼. Altera tamen pila coriacea maior & obtusiore angulo pauimentum feriens, ex altitudine pedum 37. dimissa, resultauit ad pedes proxime 6. quae dimissa ex altitudine pedum 73. resultauit ad pedes 7½. Sed ludere mihi videor, quamdiu non progredior ad experimenta nobiliora, & euidentissime demonstrantia non solum inaequalitatem in motu grauium, verum etiam incrementum velocitatis eorum vniformiter difformiter auctum versus finem motus.

VIII. *Quintum* igitur Experimentum sumptum a nobis saepissime, fuit spatij dimensio, quod graue quodpiam, aequalibus temporibus naturali descensu conficit. Id vero tentaram quidem Ferrariae cum P. Cabaeo Anno 1634. sed ex turri nostri templi non excedente centum pedes, & Perpendiculo, cuius nondum exacta Primi Mobilis tempora noram. Cum vero Bononiae essem Anno 1640. & iam diuersae altitudinis Perpendicula ex transitu Fixarum per medium caeli examinassem, minimum illud selegi pro hoc experimento, quod a vertice motus ad centrum globuli altitudinem habet vnciae vnius, & adhuc quindecim centesimas particulas vnciae pedis Romani antiqui, & quod, vt ostendi & exposui lib. 2. cap. 21. vnica sui vibratione simplici aequatur 10″. Tertijs scrupulis horarijs primi Mobilis, & senis vibrationibus Secundum vnum scrupulum exaequat exactissime; itaque vna eius vibratio simplex, est quam proxime aequalis tempore illi, quod Musici nota semichromatis designare solent, si Archimusici seu Harmostae moderantis voces eleuatio ac depressio manus ordinario modo fiat. Quia vero velocissimae atque creberrimae sunt oscillationes, seu vibrationes tam curti Perpendiculi, & ne vnius quidem errorem admittendum censui, ne in numerando confusio & fallacia vlla se oculo ingereret, post singulas decadas vibrationum, digito manus iam compressae erecto notatas, numerationem ab vnitate iterum inire soliti sumus, & contracta numerorum, qui sunt infra decadem, nomenclatura expeditissime vnum post alterum pronunciare: cuiusmodi sunt haec Italica vocabula, sed vt hic Bononiae vulgo pronunciatur breuius. *Vn, du, tri, quatr, cinq, sei, sett, ott, nou, dies.* quibus si musica, vt dixi, semichromata subijcias, & ordinariam mensuram musicae pulsationis sequaris; specimen habebis quam proximum temporum, quae vna simplex vibratio nostri huius Perpendiculi metitur. Huic porro numerationi assuefecimus & alios, praesertim PP. Franciscum Mariam Grimaldum, & Georgium Cassianum, quibus in experimento mox exponendo plurimum vius [vsus] sum. Et sane mirabile dictu est, cum binis illi & ego talibus Perpendiculis vteremur, illi quidem consistentes in coronide Turris Asinellorum, ego in pauimento subiecti peripegmatis, & in scheda vtrique seorsim sua vibrationes notassent, quibus graue inde descendebat ad pauimentum, quamuis iteratis experimentis, nunquam discrimen inter nos fuisse vnius integrae vibratiunculae. Quod scio vix creditum iri a quibusdam, & tamen verissime ita fuisse testor, & attestabuntur semper praedicti Patres e Soc. Nostra. Hactenus de Perpendiculo & mensura temporum. Praeparauimus post haec permagnum cophinum plenum globis argillaceis, quorum singuli vncias octo appendebant; & pro mensuris interuallorum breuioribus quidem vsi sumus fenestris Collegij nostri, pro altioribus autem, diuersis turribus: non quidem totis, sed fenestris aut fenestellis earum, videlicet turri S. Francisci, quae alta est pedes Romanos 150. & S. Iacobi, quae pedes 189. & S. Petronij, quae pedes 200. & S. Petri, quae pedes 208. sed praecipue & frequentius turri Asinellorum, quae tota pedes 312 .alta est, sed a coronide ad basim seu parapegma non nisi pedes 280. Quam quia commodissima est ad huiusmodi experimenta, perinde ac si ad hunc finem constructa esset, placet oculis subijcere.

IX. Esto in sequenti diagrammate [figure B.1] Turris truncus crassior IBCD, super basi fere cubica VYZX, multo crassiore quam est turris truncus, ex qua basi eminet parapegma YZH, cancellis lapideis latioribus circumseptum, vt secure circa turrim per ipsum ambulare possint sex saltem homines simul in eadem serie inter cancellos & turrim: superne autem eminet coronis BC, rostratis cancellis lapideis circumsepta, ita vt ex cancellis G, Φ, O, tamquam ex fenestris secure possit quilibet ordinariae staturae homo despicere, & perpendiculo inde demisso vsque ad pauimentum parapegmatis ID, metiri, vt nos non semel fecimus, altitudinem GI, quam, vt dixi, nacti sumus pedum Romanorum antiquorum 280. reliqua tum calamis, tum etiam totam per Altimetriam cum P. Grimaldo mensi sumus. Huiusmodi ergo Turris opportunitates maximas ad id negotium habet; nam globi ex fenestris G, Φ, O, dimissi cadunt perpendiculariter in pauimentum ID, nec impingentes in pedem turris, nec extra cancellos YZ, decidentes: deinde non est opus prohibere quemquam ne transeat per plateam basi ipsius circumstratam, interim dum globi ex coronide dimittuntur, sed absque vllius periculo possunt saepe ac saepius dimitti. Habet praeterea cingula quaedam ferrea circa F & T, cum fibulis ferreis, oppositos muros constringentibus, quibus vsi sumus pro terminis mensurandi interuallum residuum conficiendum a globo quando peruenerat in T, vel in F. Linea porro NH, supplet vicem aliarum turrium & altitudinum, quibus vsi fuimus.

X. Anno itaque 1640. mense Maio, & alijs deinde temporibus, inquisiuimus altitudinem Hβ, quae tanta esset, vt globus argillaceus vnciarum
octo sibi dimissus, percuteret pauimentum exactis praecise quinque vibrationibus perpendiculi praedicti, seu tempore 50‴. Tertiorum, & inu-
enimus eam esse pedum Romanorum 10. repetito saepius experimento.
Deinde inquisiuimus altitudinem necessariam ad descensum alterius
globi eiusdem speciei & ponderis, in duplo tempore seu vibrationibus
decem; reperimusque eam esse pedum 40. quam interuallum KH, designat. Hinc altius conscendendo, inquisiuimus altitudinem debitam triplo
longiori tempori, seu vibrationibus exacte 15. pro descensu talis globi, &
deprehendimus LH, pedum 90. Sic pro tempore vibrationum 20. inuen-
imus altitudinem MH, pedum 160. Et pro vibrationibus 25. NH, pedum
250. Tandem quia non potuimus scandere ad tantam altitudinem, quan-
tam requirunt vibrationes 30. dimissus est tamen saepe globus ex Asinel-
lae turris coronide, nempe ex G ad I, interuallo pedum 280. & percussit
pauimentum cum & ego ad I, & PP. Grimaldus ac Cassianus ad G, consistentes numerauimus vibrationes 26. vt patuit communicatis inuicem
schedis. Fingamus iam interualla linea NH, translata ad interualla lineae
OT, & primi quinarij vibrationum, seu primi temporis interuallum a
globo pertransitum esse OC, erit enim pedum 10. quantum scilicet fuit
ßH; secundi autem temporis interuallum OQ, erit aequale ipsi KH,
pedum 40. & tertij temporis interuallum OR, erit aequale ipsi LH, pedum
90. & quarti temporis interuallum OS, erit aequale spatio MH, pedum 160.
Denique quinti temporis 25. vibrationibus comprehensi interuallum totum
OT, erit aequale toti NH, pedum 250. Aduerte tamen nos hic pro T, acci-
pere signum aliquod in facie turris inter S, & pauimentum. Ergo globus
praedictus velocius ac velocius descendit, quo longius recessit ab O, &
propius accessit ad D, & separatis interuallis singulis, aequalium temporum mensuris respondentibus, primo tempore confecit OC, pedum 10.
secundo CQ, pedum 30. tertio QR, pedum 50. quarto RS, pedum 70. &
quinto ST, pedum 90. qui numeri simul conflati efficiunt 250. Quod incrementum insigne ac notatu dignum est, & sequenti experimento confir-
mandum. Hic enim solum inquirimus, an Grauia naturaliter per aerem
delabentia per rectam lineam perpendicularem horizonti, vniformiter, an
vero difformiter vniformiter, & an cum decremento, an vero cum incre-
mento velocitatis descendant.

III. Classis Experimentorum pro Proportione Incrementi
velocitatis Grauium per Aerem descendentium.

XI. E Quidem licet Anno 1629. cum primum coepi cum P. Daniele Bartolo, & D. Alphonso Iseo insigni Geometra per occasionem aliam duo perpendicula eiusdem altitudinis & ponderis simul ex eodem termino dimissa examinandi, num semper pari passu incederet quodlibet per suum arcum, aduertere oscillationes eiusdem perpendiculi esse inter se omnes ad sensum aequales in tempore seu synchronas, & cum postea Ferrariae Anno 1634. cum P. Cabaeo id ipsum certissime deprehendi, nondum intellexissem aut nouissem proportionem incrementi velocitatis Grauium a Galilaeo traditam dialogo 2. de Mundi systemate, & assertam esse secundum numeros pariter impares ab vnitate initos; immo tunc ex rudioribus meis experimentis, suspicatus essem eam esse continue triplam, videlicet iuxta hos numeros 1.3.9.27. Postea tamen facultate mihi concessa legendi dialogos illos, quos Sacra Indicis Congregatio censuris notatos vetuerat, reperi in ipsis pagina Italica 217. latina vero 163. incrementum praedictum ab eo experimentis deprehensum esse secundum numeros pariter impares ab vnitate numerabiles, videlicet vt 1. 3. 5. 7. 9. 11. &c. Suspicabar tamen in eius experimentis aliquid fallaciae latere, quia in eodem dialogo secundo pag. Italica 219. Latina 164. asserit globum ferreum centum librarum dimissum ex altitudine cubitorum centum peruenisse ad terram Secundis quinque temporis; siquidem mihi globus argillaceus vnciarum 8. descendebat ex multo maiori altitudine, videlicet ex GI, pedum 280. qui efficiunt cubitos 187. vibrationibus 26. praecise mei perpendiculi, que efficiunt tempus primi mobilis Secundorum 4″. & Tertiorum 20‴. certusque eram in mei temporis numeratione nullum sensibilem errorem fuisse, & errorem Galilaei, qui latebat in tempore non exacto ad primi Mobilis tempus & Fixarum transitum per Medium caeli, transferebam ad interualla confecta in descensu illius globi. Sed & aegre credebam potuisse illum vti globo ferreo tanti ponderis, praesertim cum nec turrim nominasset, ex qua illum demitti iussisset. Itaque suspicione hac plenus coepi, vt dixi, iam ab Anno 1640. mensuram huius incrementi tota subtilitate; sperans fore, vt aliam quandam meae fortasse propiorem inuenirem: Sed reuera deprehendi veram potius esse eam, quam Galilaeus asseruerat. Etenim ex dictis in praecedenti experimento numero 10.

exposito, agnoui illud incrementum fuisse secundum proportionem pedum 10. 30. 50. 70. 90. quod perinde est ac si in numeris minimis 1. 3. 5. 7. 9. exprimeretur. At illi nondum plane acquiescens, mutato tempore, & assumptis 6. vibrationibus perpendiculi, quae videlicet efficiunt integrum vnum Secundum primi mobilis, inquisiui cum P. Grimaldo altitudinem illis debitam, vt globus argillaceus vnciarum octo dimissus sibi, ad pauimentum peruenirent, & reperi βH, pedum 15. deinde in fine alterius Secundi, seu vibrationum praecise 12. nactus sum altitudinem KH, pedum 60. & in fine vibrationum 18. idest tertij Secundi fuit LH, pedum 135. in fine autem vibrationum 24. idest quarti Secundi, fuit MH, pedum 240. Sed pedes 280. in Turri Asinellorum iterum ac saepius pertransiuit similis globus vibrationibus 26. idest Secundis 4″. & Tertijs 20‴. Etsi enim in maioribus distantijs vnus aut alter pes vltra vel citra assumptus, non ingerat discrimen vnius integrae vibrationis, exactissime tamen praedictum numerum vibrationum, praedictis interuallis respondere deprehendi. Quare hoc pariter experimento, secernendo interualla temporibus aequalibus singillatim debita, certus fui hanc proportionem seruatam fuisse in pedibus 15. 45. 75. 105. quae prorsus est talis, qualis inter numeros 1. 3. 5. 7. Nam vt 1. ad 3. ita 15. ad 45. etc. Eamdemque reperi alijs nonnullis experimentis, quae breuitatis causa, & quia fractis numeris implicata sunt, hic praetermitto; tertium tamen aliquod expositurus in sequenti tabella [fig. B.2]. Ergo ad P. Bonauenturam Caualerium, in Bononiensi Vniuersitate primarium Matheseos Professorem, & quondam Galilaei alumnum, me contuli cum P. Grimaldo, ipsique narraui consensum meorum experimentorum cum experimentis Galilaei, quoad hanc quidem proportionem: neque enim ille chiragra simul & podagra lectulo, aut sellulae affixus, interesse ipsis poterat. Incredibile autem dictu est quantopere ex nostra hac contestatione fuerit exhilaratus.

XII. Agnoscunt iam Geometriae non ignari, spatia aequalibus temporibus compositis confecta a grauibus huiusmodi, naturaliter descendentibus, esse inter se in duplicata ratione suorum temporum; seu se habere ad inuicem vt quadrata temporum compositorum. Quod ipsum colligit *Galilaeus* dialogo illo 2. pagina eadem 217. seu 163. Nam in primo experimento, vibrationes fuere ordinatim hae: 5. 10. 15. 20. 25. quarum quadrati numeri, (idest nati ex ductu eiusdem numeri in seipsum) sunt 25. 100.

Ordo experimentorum	Vibrationes Simplices Pendiculi alti vnciam 1, $\frac{11}{100}$.	Tempus primi Mobilis respondens Vibrationibus.	Numeri Quadrati Vibrationum.	Spatia côfecta à Globo argillaceo Vnciarū 8. in fine temporū.	Spatia feorsim confecta singulis temporibus.	Proportio Incrementi Velocitatis Grauium in Aëre nostrate.
	Vibr. Simpl.	Secūda Tertia	Quadrata	Pedes Romani	Pedes Romani	Numeri minimi
I.	5	0″ 50‴	25	10	10	1
	10	1 40	100	40	30	3
	15	2 30	225	90	50	5
	20	3 20	400	160	70	7
	25	4 10	625	250	90	9
II.	6	1 0	36	15	15	1
	12	2 0	144	60	45	3
	18	3 0	324	135	75	5
	24	4 0	576	240	105	7
	26	4 20	676	280	40	8 $\frac{1}{6}$
III.	6 $\frac{1}{2}$	1 5	42	18	18	1
	13 0	2 10	169	72	54	3
	19 $\frac{1}{2}$	3 15	381	162	90	5
	26 0	4 20	676	280	118	6 $\frac{7}{128}$

Figure B.2.   Image credit: ETH-Bibliothek Zürich, Alte und Seltene Drucke.

225. 400. 625. at spatia confecta ordinatim fuere pedes 10. 40. 90. 160. 250. Iam vt 25. ad 100. ita 10. ad 40. & vt 100. ad 225. ita 40. ad 90. & vt 225. ad 400. ita 90. ad 160. & tandem vt 400. ad 625. ita 160. ad 250. Sic in 2. experimento, Vibrationum ordo fuit 6. 12. 18. 24. 26. quarum quadrata sunt 36. 144. 324. 576. 676. Interualla vero pe[r]transita & composita ordinatim fuere pedes 15. 60. 135. 240. 280. Sed vt 36. ad 144. ita 15. ad 60. & vt 144. ad 324. ita 60. ad 135. rursumque vt 324. ad 576. ita 135. ad 240. demumque vt 576. ad 676. ita 240. ad 280. Ergo spatia praedicta se habent inter se vt quadrata vibrationum seu temporum. Facilitatis tamen gratia possunt tempora illa aequalia redigi ad minimos numeros, vt primum tempus valeat 1. & secundum 2. & tertium 3. & quartum 4. & quintum 5. atque adeo adhiberi possunt quadrata haec 1. 4. 9. 16. 25. ad examinandam continue dictam proportionem. In tertio autem experimento (quod expressum habes in tabella mox exhibenda) debuerunt esse pedes 288. vt exacte praedicta proportio seruaretur, sed non licuit nobis commode id experiri, nisi ex altitudine pedum 280. Placet vero quae hactenus diximus pro tam pulchra proportione, eiusque fundamenta in vnam tabellam colligere, ac Lectori breui synopsi degustanda proponere. Addo tamen, hanc

proportionem, vt olim sperabam, non esse a nobis inuentam in pondere a globo cadente eleuabili, esto ad hanc proportionem vergat: neque enim si globus cadens ex altitudine pedis vnius, eleuat pondus vnciale, eleuabit pondus duplo maius, ex altitudine pedum 4. aut pondus quadrantale ex altitudine pedum 9. &c. vide infra classem 12. Experimentorum.

IV. Classis Experimentorum pro Duorum Grauium diuersi ponderis Descensu Inaequali ex eadem altitudine per Aerem.

XIII. Valde fallax est hoc experimentum nisi magna diligentia & circumspectione peragatur: & nisi inde conclusiones prudenter circumspectis omnibus, quae concurrere possunt ad motum incitandum aut retardandum deducantur. Comparatio enim omnium optima, & omne dubium sublatura, esset, si ex eadem altitudine, eodemque momento temporis dimitterentur duo globi eiusdem molis eiusdemque speciei, & tamen diuersi ponderis, si nimirum possemus ex duobus globis argillaceis, aut metallicis, quorum vnus esset in diametro vnius palmi & libram vnam appenderet, alter bipalmaris diametri, & libras octo appendens; palmare ita rarefacere, vt nullo ingressu alieni corporis, euaderet bipalmaris, & retineret suam soliditatem: tunc enim si grauior citius descenderet, vi solius maioris grauitatis descenderet, nec posset tarditas alterius refundi aut in diuersitatem speciei, aut in diuersitatem molis, quae ob figuram obtusiorem ac magis dilatatam difficilius aerem sub se ampliorem perrumperet: Cum in omnibus alijs esset paritas, praeterquam in grauitate. Neque vero expedit duos cylindros aut duo prismata aequalis latitudinis, sed inaequalis altitudinis ac ponderis, ex eadem materia, deorsum dimittere, fluctuant enim circa centrum grauitatis in aere, & dubiam causam celerioris aut tardioris descensus relinquunt. Sed quando hoc non licet, quod tamen huic proximum est, & coeterorum optimum medium, praeparauimus P. Grimaldus & ego duodenos globos argillaceos ex argilla solida, quorum pondus singillatim erat vnciarum 20. & alteros duodenos ex argilla illis prioribus aequales in mole, sed intus cauos, quorum singuli erant decem vnciarum, & duplo minus graues prioribus. Deinde alios multos globos diuersos in specie, vel in mole, vel in pondere collegimus, diuersamque comparationem instituimus. Et Anno 1640. mense Maio,

1645. mense Augusto, 1648. mense Octobri, ac nouissime Anno 1650. dimisimus diuersa paria ex coronide turris Asinellorum, conuocatis testibus multis e Societate nostra, qui licet non omnes semper ijsdem experimentis, interfuere, videlicet *Patres Stephanus Ghisonus, Camillus Rodengus, Iacobus Maria Pallauicimus, Franciscus Maria Grimaldus, Vincentius Maria Grimaldus, Franciscus Zenus, Paulus Casatus cum suis discipulis, Franciscus Adurnus, Octauius Rubeus,* omnes insignes ingenio, iudicio, & religiosa integritate. Et quidem inter hos aderant tres aut quatuor Philosophiae aut Theologiae Magistri, qui cum Galileo, aut Cabaeo, & Arriaga existimauerant, duo quaelibet grauia, dimissa simul ex eadem altitudine quantacumque descendere ad terram eodem physico temporis momento: At statim opinionem hanc deposuerunt. Siquidem globus argillaceus leuior seu 10. vnciarum dimissus ex G, eodem momento, quo argillaceus alter eiusdem molis sed vnciarum 20. dimissus fuit ex O; apparuit adhuc in F, distans a pauimento I, pedes saltem 15. eo momento, quo grauior pauimentum idem percusserat in D, & iam in sexcenta fragmina dissiluerat. Quamuis & ante vtriusque accessum ad pauimentum iam manifeste leuior, quem distinctionis causa ex condicto P. Grimaldus semper manu sinistra dimittebat, valde separatus a grauiore appareret paulo infra medium turris. Sed & numerato tempore ope perpendiculi grauior descendit exactis vibrationibus 23. hoc est Secundis 3″. & Tertijs 30‴. leuior autem vibrationibus 26. videlicet Secundis 4″. Tertijs 20‴. primi mobilis. Atque hoc experimentum repetitum duodecies, similem semper successum habuit. Quoniam ergo facilius obseruatu & maius discrimen erat in interualli differentia FI, quam in temporis differentia, consultum visum est, in descensu aliorum globorum, attendere potius ad residuum intervallum tardius descendentis; quam ad temporis differentiam. Quid autem nobis successerit breuius & clarius docebit sequens tabula [fig. B.3], in qua priora experimenta spectant ad binos globos eiusdem molis, posteriora ad globos diuersae molis.

Corollaria ex praecendenti Tabula.

XIV. Suppono hic Grauius in specie esse illud, quod in mole aequali naturaliter ipsi conueniente, maioris est ponderis, quomodo plumbum grauius dicitur esse cera, quia si duo globi palmares ponderentur, vnus plumbeus,

Ordo Experimentorum.	Globi eodem momento dimissi ex altitudine Pedum 180.			Distantia tardioris à paulimento ñ percussio à velociore.
	Globi eiusdem Magnitudinis	Pondus Globorū Vn- Draciæ chm.	Globus qui fuit velocior	
1	Argillace⁹ vacuus / Argillaceus solid⁹	10 0 / 20 0	Grauior	Pedes 15
2	Argillaceus vacu⁹ / Argillaceus solid⁹	20 0 / 45 0	Grauior	16
3	Argillaceus / Ligneus	9 0 / 2 4	Grauior	20
4	Argillaceus / Cereus	20 0 / 15 0	Grauior	12
5	Ligneus / Cereus	4 6 / 6 7	Grauior	15
6	Cereus / Ferreus	1 5 / 11 7	Grauior	30

Ccc 2   Al-

Altera pars Tabulæ Præcedentis.				
Ordo	Globi inequalis Magnitudinis	Pondus Globorū Vn- Draciæ chm.	Globus qui fuit velocior	Distātia &c.
7	Argillaceus / Argillaceus	5 0 / 4 0	Grauior	5
8	Argillaceus / Argillaceus	21 4 / 11 4	Grauior	12
9	Argillaceus / Argillaceus	27 3 / 14 1	Grauior	15
10	Argillaceus / Cereus	18 7 / 1 0	Grauior	35
11	Argillaceus / Plumbeus	2 0 / 1 4	Grauior	25
12	Plumbeus / Ligneus	2 4 / 2 4	Plumbeus	40
13	Argillaceus / Argillaceus	7 0 / 61 0	Grauior	10
14	Argillaceus / Argillaceus	10 4 / 23 0	Grauior	13
15	Argillaceus / Argillaceus	53 0 / 6 4	Grauior	8
16	Argillaceus / Argillaceus	53 0 / 7 1	Grauior	9
17	Ligneus Iuglādinus / Ligneus Faginus	2 1 / 4 7	Iuglādinus	2
18	Argillaceus / Plumbeus	11 0 / 1 7	Plumbeus	1
19	Argillaceus / Plumbeus	33 0 / 1 0	Argillace⁹	2
20	Argillaceus / Plumbeus	38 0 / 1 0	Argillace⁹	3
21	Plumbeus / Plumbeus	1 0 / 0 4	Grauior	5

Figure B.3. Images credit: ETH-Bibliothek Zürich, Alte und Seltene Drucke.

alter cereus, plures libras vel vncias vel drachmas etc. appendet plumbeus, quam cereus. Grauius vero in indiuiduo esse, quod maioris est ponderis absolute, siue sit eiusdem speciei, siue diuersae respectu alterius, respectu cuius dicitur grauius: quomodo globus cereus centum librarum grauior dicitur globo plumbeo vnius vnciae. Hinc consurgunt quinque combinationes vtiles. Quaedam enim dantur aeque grauia in specie & indiuiduo, vt sunt duae sphaerae plumbeae, singulae vnius librae, & hae naturaliter sunt eiusdem molis; nihil enim vetat, quin supernaturaliter aut praeternaturaliter vna illarum rarefacta maioris euadat molis. Quaedam vero sunt aeque grauia in specie, sed non indiuiduo; vt sunt duo globi plumbei, quorum vnus vnam, alter duas libras appendit, & inter haec si solida sint, id quod grauius est in indiuiduo, seu bilibre, maius quoque est, esto supernaturali aut praeternaturali addensatione possit euadere alteri aequale vel minus; at si alterum sit intus vacuum, vel plenum aere, leuioriue corpore, potest licet leuioriue in indiuiduo, aequate molem alterius. Quaedam autem sunt aeque grauia in indiuiduo, sed non in specie, cuiusmodi sunt globus plumbeus & cereus globus, quorum vterlibet sit vnius librae, & naturaliter is qui leuior est specie, vt cereus, maior est; esto supernaturaliter vel praeternaturaliter possit addensari sic vt altero non sit maior. Rursus quaedam sunt Grauiora alteris, vel in specie & indiuiduo, vt si globus plumbeus sit librarum 10. & cereus librae vnius, possuntque esse alteris aequalia in mole, vel minora, vel maiora. Quaedam autem grauiora sunt in specie, sed leuiora in indiuiduo, quomodo globulus plumbeus vnius vnciae grauior est globo cereo centum librarum; & haec sunt naturaliter minora alteris. In hac vero combinatione per correlatiuum terminum continetur, combinatio grauioris in indiuiduo sed leuioris in specie. Nam si plumbum vnius vnciae grauius quoad speciem leuius est in indiuiduo quam cera centum librarum; vicissim cera centum librarum grauior est in indiuiduo, sed leuior in specie quam plumbum vnius vnciae. His positis, corollaria infrascripta sequuntur ex praecedenti tabula.

Primo Duae sphaerae aeque graues in specie & indiuiduo, naturaliter aeque velociter per idem medium descendunt; intellige semper, si simul ex eadem altitudine dimittantur: sunt enim aequalis molis, & ita omnia sunt paria. Si tamen supernaturaliter aut praeternaturaliter vna ex illis aut rarefactione maior, aut condensatione minor euaderet, minor velocius descenderet, quia angulus contactus, sed interior, quo planum aeris subjecti feriret, esset acutior.

Secundo Duarum sphaerarum aeque grauium in specie, sed non in indiuiduo, illa naturaliter citius descendit per aerem, quae grauior est in indiuiduo; siue sint aequalis molis, quia vna earum sit intus caua, vt patet experimento 1. duodecies repetito, & experimento 2. siue diuersae molis, vt patet experimentis 7. 8. 9. 13. 14. 15. 16. 21. Ex quibus aliqua sunt, in quibus crediderim non seruatam proportionem in velocitate cum excessu in grauitate indiuidua, similem ei, quae in alijs apparuit, vel ob magnitudinis excessum nimium, qui angulum obtusum efficeret; vel quia globi non vsquequaque perfecte rotundati essent: sed praestat experimenta fideliter referre potius, quam illa limitando concinnate.

Tertio Duarum sphaerarum aeque grauium in indiuiduo, sed non in specie, illa naturaliter citius descendit per aerem, quae in specie grauior est. Quod constat experimento 12. Et ratio est, quia naturaliter illa quae grauior est in specie, minoris est molis altera sibi in indiuiduo aeque graui, quare illam superat non solum grauitate specifica, sed etiam acumine figurae seu anguli contactus, ex paruitate sphaerae orti: praeterquamquod in casu nostro, lignum porosius est plumbo, & ob aeris vel halituum intra poros latentium leuitatem minus idoneum ad grauitandum.

Quarto Duarum sphaerarum, quarum vna est grauior tum in specie, tum in indiuiduo, illa naturaliter citius descendit per aerem, quae grauior est: siue sit maior altera, vt patuit experimento 10. siue illi aequalis, vt constitit experimentis 3. 4. 5. 6. siue etiam minor, vt vidimus experimento 11.

Quinto Duarum sphaerarum, quarum vna grauior est in specie, sed non in indiuiduo, illa quae grauior est in specie, potest citius (naturaliter loquendo) per aerem descendere; potest & aeque velociter; potest & tardius. Primi casus exemplum habuimus experimentis 12. 17. & 18. Tertij vero casus exemplum in experimentis 19. & 20. Secundi autem & tertij habemus argumentum sufficiens conferendo experimentum 8. cum 18. Etenim in vtroque Argillaceus globus fuit 11. vnciarum, neglectis interim drachmis; atqui argillaceus vnciarum 11. dimissus cum argillaceo vnciarum 21. experimento 8. ita tardior fuit, vt distaret pedes 12. a pauimento, quando argillaceus vnciarum 21. illud percussit: argillaceus autem item vnciarum 11. experimento 18. dimissus cum plumbeo vnciarum fere 2. non fuit tardior nisi quantum designat vnius pedis interuallum: Ergo si cum plumbeo vnciarum fere 2. dimissus fuisset argillaceus 21. vnciarum, reliquisset post se plumbeum interuallo pedum 11. aut saltem plurium quam vnius, & euasisset velocior in descensu: cum ergo aucta grauitate indiuiduali euaserit de tardiore velocior, vtique potuit tam parum augeri grauitas argillae indiuidualis, vt aeque velociter cum plumbo descenderet. Dixi *saltem plurium,* mouet enim me experimentum 19. & 20. in quibus plumbeus globulus vnius vnciae ob paruitatem plus acquisiuit velocitatis, quam defecerit a grauiore argillaceo ratione ponderis, licet longe maioris, adeo vt tardior quidem fuerit argillaceo, sed paululo tardior: ad quod etiam contulit rotunditas perfectior plumbeorum, quam argillaceorum.

Sexto Duarum sphaerarum simul ex eadem altitudine per eumdem aerem dimissarum, nunquam illa naturaliter citius aut aeque velociter descendit, quae vndecumque leuior est: aut enim est grauior in specie & indiuiduo, aut grauior in indiuiduo, aut grauior in specie. Quare simpliciter loquendo, quae grauiora sunt citius descendunt naturaliter, in nostro quidem aere. Raro autem euenit vt sint diuersi ponderis, & defectus grauitatis indiuiduae compensetur acumine paruitatis in mole, vt aeque velociter descendant.

PART 3
Technical Discussion of Riccioli's Reports Regarding
His Experiments with Falling Bodies

Riccioli's reports regarding his falling bodies experiments are clear and straightforward from a technical standpoint. Only on a few points do they benefit from further discussion.

Riccioli is writing in a time when the most basic behaviors of falling bodies could still be disputed. Thus he begins in his discussion of "Unequal motion of Heavy Bodies" by establishing basic behaviors of falling bodies. In his first few paragraphs he illustrates simple experiments that show that the impetus (momentum) gained by a falling body increases with the distance traversed in the fall, and that it is not dependent in an absolute way on the height of the point of release. The sound made by a ball dropping into a bowl or by water dropping into a container, the impact of a ball dropped into a hand, or the bounce of a ball dropped from a height, all illustrate the point Riccioli is trying to make. So do the images from literature that he calls upon.

Having illustrated basic behavior, Riccioli next moves to providing a detailed description of the procedure used in the experiments on "the Proportion of the Growth of the Velocity of Heavy Objects descending through the Air." He shows how he and Grimaldi measured time and height, and how they minimized and tried to quantify errors in these measurements. He also describes the balls that they dropped—clay balls weighing eight ounces.

Next comes a discussion of how Riccioli and Grimaldi determined the distance a ball travels in falling for a time of five, ten, and so on strokes of his timing pendulum. It appears that they first determined the distance a ball drops during five pendulum strokes. Then they tested whether the distances dropped in multiples of those five strokes followed a progression of square numbers as described by Galileo in his *Dialogue* (the distances did, in fact, follow that progression). The method Riccioli and Grimaldi used suggests that, of the several measurements given, only the initial five-stroke (ten-foot) drop appears to be a fully independent measurement. Riccioli's further discussions and his table suggest that they followed this same method with a six-stroke drop (followed by multiples) and a six and

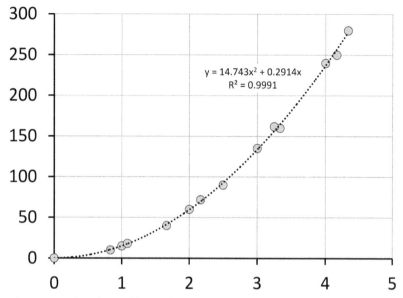

Figure B.4.   Plot of Riccioli's data from table B.1. Vertical axis is distance in Roman feet. Horizontal axis is time in seconds.

one-half–stroke drop (followed by multiples). Thus arguably their only fully independent measurements are the five-stroke drop, the six-stroke drop, the six and one-half–stroke drop, and the final 280-foot drop.

Figure B.4 is a plot of all of Riccioli's measurements. The curve fit to these data indicates that his falling clay balls gained 29.5 Roman feet per second during each second of fall—an "acceleration due to gravity," or g, of 29.5 Roman feet per second per second (Rmft/s^2). The four fully independent measurements are plotted in figure B.5. The fit to these data yields g = 29.8 Rmft/s^2. Calculating g using the four independent measurements and the equation d = ½gt^2 (the equation for the distance d that a falling body moving with acceleration g travels in time t) yields an average value for g of 29.8 Rmft/s^2 with a standard deviation of 0.7 Rmft/s^2.

If Riccioli's Roman foot is that measurement commonly given today as 0.296 meters, then the modern accepted value of g found in basic physics texts (9.8 meters per second per second, or 9.8 m/s^2) would be 33.11 Rmft/s^2. Thus Riccioli's value of 29.8 Rmft/s^2 is off by about 10 percent from the accepted value. This means that for the 280–Roman foot

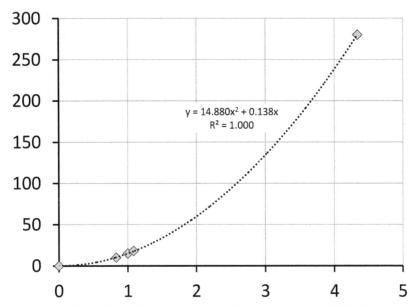

Figure B.5.   Plot of Riccioli's four fully independent measurements, highlighted in table B.1. Vertical axis is distance in Roman feet. Horizontal axis is time in seconds.

drop, Riccioli "should" have measured a time of 24.7 strokes rather than the 26 strokes that he did measure, or 4.11 rather than 4.33 seconds. Yet Riccioli is certain that his times are not off by more than a stroke. Moreover his times are consistently high (his five-stroke drop "should" be 4.7 strokes; his six should be 5.7; his 6.5 should be 6.3), and his errors do not increase in a manner obviously consistent with the effects of air resistance. Thus he may well have a systematic error in distance measurement.

Alternatively, his Roman foot simply may not be 0.296 m. One way to determine the size of Riccioli's Roman foot is to use his statement that the Asinelli tower is 312 feet tall. Modern measurements give its height as 97.38 m,[12] so this suggests that Riccioli's Roman foot is 0.312 m. If this is the case, then the accepted value of g would be 31.40 Rmft/s^2; Riccioli's value is off by about 5 percent from the accepted value; Riccioli "should" have measured a time of 25.3 strokes rather than 26 strokes he did measure for the 280 Rmft drop (within his stated margin of error); and the other drop times are even closer to the values they "should" have.[13] At any rate, Riccioli's results are quite good. Many physics students equipped with modern knowledge of physics and precision equipment made especially for the purposes of determining the value of g have been happy to measure it to within 10 percent accuracy.

As the modern reader moves on to the paragraphs that bring the "Proportion of the Growth of the Velocity" section to a close, he or she will doubtlessly have a great appreciation for the ease, convenience, and brevity of today's mathematical notation! Statements like "as 25 is to 100, so 10 is to 40; as 100 is to 225, so 40 is to 90; as 225 is to 400, so 90 is to 160" and so forth are a tedious way to convey mathematical concepts.

The final paragraph of this section contains comments on the amount that a falling ball that strikes one pan of a balance will elevate a weight in the other pan. This correlates to what would be considered the energy of the ball today. Riccioli had expected this value to proceed as the square of the height from which the ball falls. However, the gravitational potential energy (and thus the kinetic energy upon impact) of a falling object increases directly with height. We have not translated the experiments to which Riccioli refers at the very end of the "Proportion of the Growth" section.

Finally comes the section on the "Unequal Descent of Two Heavy Bodies of unlike weight through the Air." In modern terms, these are experiments on the effect of air resistance or air drag on falling bodies. Riccioli's ultimate conclusion is that heavier bodies do fall to the ground more rapidly than lighter ones—if by "heavier" one means bodies that are either heavier individually, or heavier specifically[14] (that is, denser). However, Riccioli expresses surprise to find that air can have a significant effect. Tycho Brahe had believed that air was too tenuous to have any significant effect on a heavy projectile (see chapter 2).

Standard textbook physics says that the acceleration a of a spherical body (a ball) of mass m, density ρ, radius r, cross-sectional area $A = \pi r^2$, and drag coefficient C_d, falling through air of density ρ_{air} under the influence of its weight ($W = mg$) and air drag ($D = \frac{1}{2}\rho_{air}AC_dv^2$) is given by:

$$a = dv/dt = F/m = (W - D)/m = (mg - \frac{1}{2}\rho_{air}AC_dv^2)/m$$
$$= g - (\frac{1}{2}\rho_{air}\pi r^2 C_d v^2)/(4\rho\pi r^3/3)$$

or

$$a = g - (3\rho_{air}v^2/8)(C_d/\rho r)$$

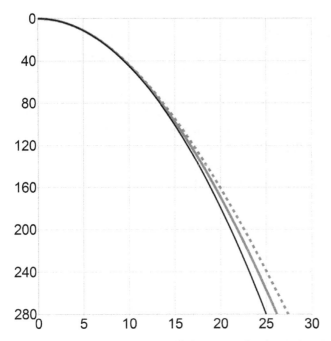

Figure B.6.   Plots of distance travelled in Roman feet (vertical axis) vs. time in pendulum strokes (horizontal axis) for two different balls (gray solid line and gray dashed line) falling through air. Also shown is a plot for a ball falling with negligible air drag (solid black line). These plots were generated via a computer simulation of a falling body with air resistance. The solid gray line is a 20-ounce ball of diameter 10 cm. The dashed gray line is a 10-ounce ball of the same diameter. These are similar to the solid and hollow clay balls that Riccioli dropped. Riccioli reports that the lighter ball was about 15 feet from the ground when the heavier ball struck the ground, which is generally consistent with the simulation. Denser, smaller bodies do reach the ground more rapidly, as Riccioli discovered.

Thus the ball's acceleration will be measurably less than g if the ball is moving with sufficient speed v. Also, the lesser the ball's density ρ, or radius r, and the greater its drag coefficient C_d, the lesser the ball's acceleration, whatever the speed, and the longer the time required for it to drop from a given height (fig. B.6).

Riccioli — recognizing that the source of the different fall times of different balls is the interaction of the balls with the air (he also experimented with bodies falling through water; these experiments come right after "Unequal Descent of Two Heavy Bodies of unlike weight through the Air")—reaches these same conclusions through his experiments. He explicitly recognizes that lower density ρ, or "specific heaviness," and radius r, or size or "bulk" (volume), reduces acceleration. As regards drag coefficient C_d, he simply notes that balls that are less perfectly round are prone to more air drag.

Riccioli's experiments with different falling bodies successfully identified the various factors that determine the effect of air resistance on a falling body. This along with his careful experiments to determine how the speed of a falling body changes over time provided a thorough description of the motion of a falling body—a description published in a prominent work.

Notes

Chapter 1. Giovanni Battista Riccioli and the New Almagest

1. Riccioli described Jupiter as having "fascia"—bands, or streaks of cloud—on its disk. See Graney 2010c, 265.

2. English translations of biblical verses in this volume are from the Douay-Rheims version, which would have been available in Riccioli's time.

3. Dinis 2003, 195–98.

4. Riccioli 1651, 2:478.

5. Riccioli 1651, 2:478: "Spectata sola Ratione & Argumentis Intrinsecis, et omni Auctoritate circumscripta; Absolute asserenda est tanquam vera Hypothesis Immobilitatem seu Quietem Terrae supponens; & falsa ac demonstrationibus Physicis imo et Physicomathematicis repugnans illa Hypothesis, quae terrae vel solum Diurnum, vel Diurnum et Annuum motum tribuit." Also see Dinis 2003, 209, although Dinis misreads and is critical of Riccioli's reasoning.

6. Riccioli 1651, 2:496–500. The condemnation by church authorities that is referred to here is the 5 March 1616 decree by the Congregation of the Index that declared the Copernican hypothesis to be "false" and "altogether contrary to Holy Scripture." See Finocchiaro 1989, 30, 149; Riccioli 1651, 2:496.

7. Grant 1996, 652.

8. Riccioli 1651, 2:193–536 (Book 9). For a complete listing of all 126 arguments in English, with brief discussions of each, see Graney 2011d.

9. For an example of such an argument on the heliocentrism side, see Riccioli 1651, 2:468 (par. 15). For an example on the geocentrism side, see 2:476 (par. 62). This will be discussed further in chapter 7.

10. Riccioli 1651, 2:469 (par. 22): "Sol est centrum systematis Planetarij, cum circa illud iam certum sit moueri Mercurium ac Venerem, vnde de alijs similis sit coniectura. Ergo debet esse centrum Vniuersi."

11. Riccioli 1651, 2:469 (par. 22, *Responsum*).

12. Riccioli 1651, 2:475–76 (par. 42 and *Responsum*).

13. Riccioli generally did not use inclusive language.

14. Riccioli 1651, 2:476 (par. 53 and *Responsum*).

15. Riccioli 1651, 2:477 (pars. 62–70 and *Respondent*).

16. Riccioli 1651, 2:467 (par. 9 and *Responsum*).

17. Riccioli 1651, 2:474 (par. 17 and *Respondent*).

18. Riccioli 1651, 2:467 (par. 11 and *Responsum*).

19. Riccioli 1651, 2:477 (par. 70 and *Respondent*): "etsi falsitatis redargui non possit; prudentioribus tamen viris non posse satisfacere."

20. Galilei 2001, back cover.

Chapter 2. The Universe that Riccioli Saw

1. Keill 1739, 47–48.

2. "New Stars" in Hill 1754.

3. Abrams 1962, 1293–94.

4. Aristotle's *Physics* (bk. 8, part 10) in Apostle and Peterson 1986, 237.

5. Apostle and Peterson 1986, 238.

6. Buridan 1974, 275–76.

7. Buridan 1974, 276–77.

8. Buridan 1974, 277.

9. For a discussion of Buridan's ideas and physics, see Graney 2013c.

Chapter 3. The Anti-Copernican Astronomer

1. Couper, Henbest, and Clarke 2007, 120; Thoren 1990, 188.

2. I thank Owen Gingerich for this discussion.

3. Maeyama 2002, 118–19.

4. Gingerich 1973, 87.

5. Gingerich 2009, 10:00 mark and following.

6. Gingerich 1993, 181.

7. Blair 1990, 361.

8. There were various versions of hybrid geocentrism. Ones that were similar to Tycho's but not exactly the same (the version that appears in Riccioli's frontis-

piece differs slightly from Tycho's, for example) are technically "semi-Tychonic" world systems. See Schofield 1989. For the purposes of this book these will all be considered as minor variations on the same idea.

9. See Jarrell 1989, 29.

10. Gingerich and Voelkel 1998, 1.

11. Copernicus (1543) 2001, 133.

12. Van Helden 1985, 27, 30, 32, 50.

13. See Thoren 1990, 302–4.

14. Thoren 1990, 304–6.

15. Blair 1990, 364; Moesgaard 1972, 51; Brahe 1601, 167, 191.

16. Van Helden 1985, 51.

17. I borrow this phrase from Dava Sobel's book of the same title.

18. Brahe 1601, 189: "is uiolentissimus motus alteram, quo grauia necessario, et naturaliter recta descendunt, adeo impediat, ut nisi post longe emensum spatium, imo uix quidem antequam uiolentia illa se remiserit, atque; in quietem paulatim desierit, Terram contingere possit"

19. Brahe describes the air as "liquidissimo aere," "Aer qui adeo tenuis est," and "Aere etiam tam tenui"; Brahe 1601, 188–89.

20. See Graney 2013c, 412–13.

21. Brahe 1601, 188–90.

22. Galilei 2001, 216–18.

23. Galilei 2001, 218.

24. Brahe 1601, 189–90. "Addo uero et hoc, quod si circa Terrae polos, ubi motus diurnus (si quis esset) in quietem desinit, eadem fieret uersus quamcunque Horizontis partem per sclopetum ratione ante dicta experimentatio, idem omnimode eueniret, ac si in medio inter utrumque polum apud Aequatorem, ubi motio circumferentiae Terrae concitatissima esse deberet: Vti etiam in quouis Horizonte, si uersus ortum et occasum parili ratione emittatur globus, idem conficit spatij, quod uersus Meridem et Septentrionem simili impulsione emissus, cum tamen Terrae, si quis inesset diurnus motus, is occasum ortumque respiceret: Meridiem uero et Septentrionem non item: Cum igitur haec uniformiter ubique; eueniant, quiescat etiam ubique; uniformiter Terra, necessum est."

25. Galilei 2001, xxiii, xxviii.

26. "Tycho Brahe's verdensbillede" 2013.

27. For example, Wertheim 2003.

28. For example, Freedberg 2002, 83.

29. See the encyclopedia article quoted in the epigraph at the start of this chapter. Christiaan Huygens refers to Tycho's "principal argument" in his *The Celestial Worlds Discover'd* (1722, 145). I thank Dennis Danielson for bringing Huygens's comment to my attention.

Chapter 4. Stars and Adventitious Rays

1. Drake 1957, 47, note 16.
2. Schofield 1989, 41.
3. Drake 1957, 47.
4. Kepler 1995, 46.
5. Van Helden 1985, 89.
6. Grant 1996, 448.
7. Drake 1957, 100.
8. Drake 1957, 137.
9. Ondra 2004, 73; Favaro 1890–1909, vol. 3, part 2, 877.
10. Galilei 1989, 167.
11. Galilei 1989, 174.
12. Galilei 2001, 417.
13. Galilei 2001, 451.
14. Galilei 1989, 167.
15. Galilei 2001, 417.
16. Ondra 2004, 73; Favaro 1890–1909, vol. 3, part 2, 877.
17. Galilei 2001, 444. Also see Galilei 1989, 176.
18. Galilei 2001, 444.
19. Graney 2007.
20. Consider Galileo's observations of the double star Mizar as an example. As noted earlier in this chapter, Galileo recorded that the two component stars of Mizar were separated by 15 arc seconds, and that one component star (which we shall call Mizar A) measured 3 seconds in radius, and the other (Mizar B) measured 2 seconds. These radii correspond to apparent diameters of 6 and 4 seconds, respectively. The apparent diameter of the Sun is approximately 1800 seconds, which is 300 times greater than Mizar A and 450 times greater than Mizar B. Galileo, assuming these stars to be suns, calculated Mizar A to thus be 300 times more distant than the Sun (Ondra 2004, 73; Favaro 1890–1909, vol. 3, part 2, 877). The parallax angle of Mizar A, determined by the ratio 1/300, is 0.00333 radians, or 688 arc seconds. The parallax angle for Mizar B, determined by 1/450, is 458 seconds. The difference between these, 230 seconds, is the differential parallax of this double star. The separation between Mizars A and B should vary annually by roughly this much, owing to Earth's motion around the Sun. Granted the 15 second separation of Mizars A and B, and Galileo's assumptions about these stars being suns at distances several hundred times greater than the Sun, these stars should manifest a dramatic change in separation, caused by Earth's motion, in relatively little time. This in fact does not happen.
21. For a full technical discussion of this problem, see Graney 2008; 2010a.

22. Siebert 2005, 254–56.

23. Siebert 2005, 254–60; Ondra 2004, 73–74.

24. Graney and Sipes 2009.

25. Ondra 2004.

26. Christianson 2000, 319–21.

27. Watson 2005, 85–86; Bond 1848, 75–76.

28. Marius 1614, fifth page of "Praefatio ad candidem lectorem": "Inter illa primum est, quod mediante perspicillo a die 15. Decemb. Anni 1612. invenerim et viderim fixam vel stellam quandam admirandae figurae, qualem in tota coelo deprehendere non possum. Ea autem est prope tertiam et borealiorem in cingulo Andromedae. Absque instrumento cernitur ibidem quaedam quasi nubecula: at cum instrumento nullae videntur stellae distinctae, ut in nebulosa cancri et alijs stellis nebulosis, sed saltem radij albicantes, qui quo propiores sunt centro eo clariores evadunt, in centro est lumen obtusum et pallidum, in diametro quartam fere gradus partem occupat. Similis fere splendor apparet, si a longinquo candela ardens per cornu pellucidum de nocte cernatur; non absimilis esse videtur Cometae illi . . . Anno 1586."

29. Pannekoek 1961, 231.

30. Dreyer 1909, 191.

31. Marius 1916. The missing material, which also includes his notes on Andromeda, would be located after the end of the Preface on page 373 and before the beginning of part 1.

32. Marius also claims the Tychonic system as his own.

33. Marius 1614, sixth and seventh pages of "Praefatio ad candidem lectorem": "quod non ita pridem, videlicet post reditum a Ratisbona mihi pararim instrumentum, quo non solum planetae, sed etiam, omnes fixae insigniores exquisitae rotundae cernuntur, inprimis autem canis major, minor, lucidiores in Orione, Leone, Vrsamajore, etc. quod antehac nunquam mihi videre contigit. Miror equidem Galilaeum cum suo instrumento admodum excellente idem non vidisse. Scribit enim in suo sidereo Nuncio, fixas stellas periphaeria circulari nequaquam terminatas apparere, id quod quidam postea maximi argumenti loco habuerunt, nimirum hoc ipso systema mundanum Copernicanum confirmari, nempe quod ob immensam distantiam fixarum a terra, figura globosa fixarum stellarum nequaquam in terris ullo modo percipi possit. Cum vero nunc certissime constet, etiam fixas orbiculari in terris hoc perspicillo videri, cadit profecto haec argumentatio, et plane contrarium astruitur, nimirum sphaeram stellarum fixarum nequaquam adeo incredibili distantia a terris removeri, uti fert speculatio Copernici, sed potius talem esse segregationem sphaera fixarum a terris, ut nihilominus moles corporum illarum hoc instrumento figura circulari distincte videri possit, consentiente ordinatione sphaerarum coelestium, Tychonica et propria"; Graney 2010a, 18.

34. Marius 1916, 404, 408–9.

35. Marius 1614, seventh page of "Praefatio ad candidem lectorem"; Graney 2010a, 18.

36. Crüger 1631, Sig. Jj ir. I thank Dennis Danielson for this quote, as he discovered it, translated it from German, and brought it to my attention.

37. Van Helden 1985, 89.

38. An interesting question that arises is this: What would have been the effect on the world system debate had Galileo published his double star observations?

39. Graney 2010b, 462.

40. Graney 2009.

41. Drake 1957, 46.

42. Drake 1957, 46.

43. Galilei 2001, 88, 390.

44. Galilei 2001, 418–19.

45. Riccioli 1651, 1:715–16: "Tubo denudanti discos stellarum, et abradenti cincinnos radiorum aduentitios"; Graney 2010b, 458.

46. Graney 2010c.

47. Drake 1957, 100.

48. Drake 1957, 137.

Chapter 5. Science against Copernicus, God's Starry Armies for Copernicus

1. Translation of the consultants' statement of appraisal regarding the proposition that the Sun is immobile: "Omnes dixerunt dictam propositionem esse stultam et absurdam in Philosophia; et formaliter haereticam, quatenus contradicit expresse sententiis sacrae scripturae in multis locis, secundum proprietatem verborum, et secundum communem expositionem, et sensum, Sanctorum Patrum et Theologorum doctorum." I have purposefully retained the punctuation and capitalization found in the original manuscript at the Vatican Secret Archives. Finocchiaro (1989, 344, note 35) notes the presence of the semicolon after "Philosophia." He also notes that some secondary sources show a comma, while others have no punctuation. The latter case, he says, "conveys the impression that biblical contradiction is being given as a reason for ascribing both philosophical-scientific falsehood and theological heresy." A review of secondary sources reveals that few show a semicolon after "Philosophia," while many show a comma or no punctuation. Sources that show the semicolon include von Gebler (1877, 47–48), Costanzi (1897, 227), and Grisar (1882, 38). Sources that show a comma include the 2009 volume by Pagano, from the Vatican Secret Archives (42–43), and Favaro (1890–1909, 19:321). This suggests possible disagreement over what punctuation

follows "Philosophia" (however, Finocchiaro, at the same note 35, points out a 1984 version of Pagano as showing no punctuation whatsoever). An image of the original document shows that the upper mark in the semicolon is elongated, so perhaps the semicolon has been taken to be a stray mark and a comma. However, the consultants' next statement of appraisal, regarding the proposition that the Earth has motion, reads: "Omnes dixerunt, hanc propositionem recipere eandem censuram in Philosophia; et spectando veritatem Theologicam, adminus esse infide erroneam [All have said, this proposition to receive the same appraisal in Philosophy; and regarding Theological truth, to be at least erroneous in faith]." The semicolon following "Philosophy" in this second statement is clear, and appears in Pagano, Favaro, and other secondary sources. Parallel structure between the two statements supports the presence in the first statement of the semicolon after "Philosophy." See Graney 2014, which includes images of the original document from 24 February 1616.

2. See Donahue 1981, 103; Galilei 2001, 105, 270, 414, 426, 551, 571; Finocchiaro 2014, 139–49, 194–97.

3. Locher 1614, 28: "Argumenti Nucleus" 1.

4. Locher 1614, 23–25: "In hac itaque partium Vniuersi collocatione et commentitia terrae motae hypothesi, Sol, Mercurius et Venus infra, terra est supra: et grauia absolute et naturaliter ascendunt, leuia descendunt; et Christus Dominus ascendit ad infernos, descendit ad coelos, quando Soli appropinquauit; et quando Iosue Solem iussit stare, vel terra stetit, vel Sol motus est contrarie terrae: et quando Sol in Cancro versatur, terra Capricornum percurrit, atque vniuersim hyemalia signa aestatem, aestiualia hyemem creant: sidera non terrae, sed terra sideribus oritur occiditque: ortus in occasu, occasus in ortu incipit; et denique totus quasi Mundi cursus inuertitur." See also Galilei 2001, 415.

5. Locher 1614, 25: "Et ad haec similiaque, licet torte satis respondeant Copernicani."

6. Locher 1614, 25: "minus tamen poterunt satisfacere his quae sequuntur."

7. Locher 1614, 28: "Argumenti Nucleus" 2.

8. Locher 1614, 26.

9. Locher 1614, 28: "Argumenti Nucleus" 3; 35: "Nervus Argumenti" 1–3.

10. Locher 1614, 50–53.

11. Fantoli 1994, 240–41; Finocchiaro 2010, 72; Finocchiaro 1989, 347, note 2.

12. Finocchiaro 2010, 72.

13. Finocchiaro 1989, 155.

14. Finocchiaro 1989, 155.

15. Finocchiaro 1989, 28–29, 134–46.

16. Finocchiaro 1989, 29–31, 146–50.

17. Favaro 1890–1909, 5:412. See appendix A, part 1, chap. 7 (175 in this volume).

18. Finocchiaro 1989, 154–97.

19. Favaro 1890–1909, 5:405–6. See appendix A, part 1, chap. 2 (167 in this volume).

20. Favaro 1890–1909, 5:406. The Latin for this is provided in n. 23 below.

21. Finocchiaro 1989, 167, 174.

22. Finocchiaro 1989, 172.

23. Favaro 1890–1909, 5:406: "quae distantia adeo magna non solum asymmetrum esse universum ostendit, sed etiam convincit, aut stellas fixas nihil operari posse in haec inferiora, propter nimiam earum distantiam . . . ; aut stellas fixas tantae magnitudinis esse, ut superent aut aequent magnitudinem ipsius circuli deferentis Terram." Ingoli's parenthetical comment about the Sun, before the semicolon, is omitted.

24. Favaro 1890–1909, 5:408. See appendix A, part 1, chap. 5 (171 in this volume).

25. Graney 2012a, 113.

26. Finocchiaro 1989, 68; Graney 2011a, 70.

27. Finocchiaro 2005, 285–86. Bellarmine, in lectures given in 1571 at Louvain University, discussed the question of whether the heavens were solid (so that the fixed stars were attached to the celestial sphere, whose rotation carried them around the Earth as a unit) or whether they were fluid (so that each star revolved around the earth individually). Bellarmine supported a fluid heavens, mostly on biblical grounds. However, he said, "But if in the future it will be conclusively proved that the stars move by the motion of the whole sky and not on their own, then one would have to see how to understand the Scriptures so that they do not conflict with an established truth. In fact, it is certain that the true meaning of Scripture cannot conflict with any other philosophical or astronomical truth."

28. Finocchiaro 2005, 93.

29. Finocchiaro 2005, 93–94.

30. 1 Kings 7:23.

31. Revelation 7:1 and Mark 4:31, respectively.

32. Copernicus (1543) 2001, 133.

33. Digges 1573, "Operis Conclusio" (last paragraph): "a diuino illo plusquam humani ingenij Copernico." Also see Graney 2012a, 100–102.

34. Digges 1573, "Operis Conclusio" (especially last paragraph); also see Graney 2012a, 100–102. I have translated, paraphrased, and condensed Digges's words, although the last sentence is a fairly close translation of the last sentence of Digges's *Alæ*: "indubieque alijs stupendum DEI Miraculum testificari, quibus non datum est a Terris sursum attollere vultus, vt cunctis denique; innotescant mag-

nalia DEI, cui soli omnis LAVS, HONOR, et GLORIA, exhibeatur in Aeuum." For a discussion of Digges's *Alæ*, see Goulding 2006.

35. Johnson and Larkey 1934, 78.

36. See Johnson and Larkey 1934, 78. However, I have opted to quote a version with somewhat modernized spelling found in Danielson 2000, 137.

37. Brahe 1601, 186: "cur mihi non verisimile videatur, spacium a Sole ad Saturnum tot vicibus contineri intra Saturnum et affixarum Stellarum remotionem? aut quid absurdi sequitur, si Stella tertiae Magnitudinis aequat totum orbum annuum? An id aut cum voluntate diuina pugnat, aut diuinae Naturae impossibile est, aut infinitae Naturae non competit? Haec demonstranda omnino tibi sunt, si absurdi quid hinc colligere volueris. Non tam facile absurditatis argui possunt, quae vulgo absurda prima fronte videntur, sed longe maior est diuina Sapientia et Maiestas. Et quantumcunque; etiam Mundi Vastitatem, et Magnitudinem concedas, nullam tamen proportionem ad infinitum Creatorem habebit. Quo maior Rex, eo maius et amplius palatium decere putat suam Maiestatem. Quid cogitabis de DEO?"; also see Graney 2012b, 217.

38. Brahe 1601, 191–92: "Ad quid hoc est dicere? Num uoluntas Diuina irregulariter et in ordinate uspiam agit, contra, quam alias in toto Mundi Theatro apparet? Vbi omnia iusto tempore, mensura et pondere undiquaque rite disposita sunt? Nihil uacui, nihil frustranei, nihil sibi inuicem non certa harmonia, et proportione correspondens. Scilicet a Saturno ad fixas Stellas non quidpiam in usus Terricolarum destinatum continebitur, per interuallum, plusquam 700000. Semidiametrorum Terrae, Stellis fixis quae longe superius elatae sunt, his tamen non minimum inseruientibus. Scilicet etiam fixa sidera nonnulla totum Orbem Annuum, quem Sol describit (siue, ut ille uult, Terra) sua magnitudine aequabunt, nonnulla uero adhuc longe maiora erunt. Ipse vero Sol, praecipuum Mundi corpus ac luminare huic quantitati uix conferendus uideatur, cum tamen Stellae in Coeli expanso constitutae, minimam, respectu Solis in patefactione Creationis Mundi obtineant praerogatiuam, uti etiam per se comparatione eius quam minimam habent; Sed quasi ab Authoritate huius, et praeminentia, uti et reliqui Planetae dependeant. Et qualis quaeso Mundi uisibilis Symmetria sic prodibit, si maxima pars Creaturis uisibilibus destituta erit: Quaedam uero corpora coelestia in immensum ferme augeantur, quaedam uero utut per se uasta, cum his tamen uix conferri possint?"; Graney 2013a, 166.

39. Brahe 1601, 192; Graney 2013a, 167.

40. Barker 2004.

41. Jorink 2010, 52; Howell 2002, 146–47; Vermij 2002, 81–84.

42. Vermij 2002, 89.

43. Lansbergen 1663, 27: "DE VERA DELINEATIONE SECUNDI COELI"; Graney 2013a, 168.

44. Lansbergen 1663, 27: "Vnde sequitur, singulas Fixas, Sphaeram Terrae magnitudine aequare; quod et captum, et fidem humanam, paene superat."

45. Lansbergen 1663, 27–28: "Addo, quod Proprietatibus *Dei* melius conveniat haec Sphaerae *Fixarum* Magnitudo, quam *Tychonica*: quia ex ea rectius percipitur, *Deum* immensum esse atque infinitum. Nam *Secundum Caelum* infinito simile, quid arguat? nisi *Deum* revera infinitum; cum *Coelum ac Terram impleat*, Ier. 23. 25. et *Coelum Coelorum ipsum non capiat*, 1 Reg. 8. 27. Gloriosius denique supremae Majestati *Divinae;* quod tam vastum *Atrium, Palatio* suo praestruxerit, quam si id minus fecisset"; Graney 2013a, 168. I have changed Lansbergen's scriptural references to agree with modern versions.

46. Lansbergen 1663, 28: "Itaque rationem habuit, *Numeri, Mensurae,* ac *Ponderis,* in utriusque Caeli constructione: eosque absoluta perfectione inter se coadunavit"; Graney 2013a, 168.

47. Lansbergen 1663, 28: "Haec necesse est recte percipiamus; ne stupendae Caelorum, Corporumque Caelestium magnitudini inhaerentes; ea habeamus loco Creatoris, ut quondam *Ethnici*: sed solum *Deum* honoremus ac veneremur, ut *Creatorem laudandum in omnia secula, amen.* Rom. 1. 25"; Graney 2013a, 168.

48. Lansbergen 1663, 28: "licet ingens sit spatium, intra Orbem *Saturni,* et *Fixarum*; non tamen est vacuum, ut *Tycho Brahe* cum asseclis, sensit; sed Creaturis *Dei* repletum. . . . tanta multitudine Creaturarum invisibilium undique obsidetur."

49. Lansbergen 1663, 28: "sequitur, *Secundum Coelum* non esse vacuum, sed Spiritualibus his Creaturis, undique repletum."

50. Lansbergen 1663, 29: "Hae igitur Scalae *Iacobo* visae, uti nusquam vacuae erant, sed *Bonis* Angelis undique plenae; Ita *Secundum Coelum* nunquam vacuum, sed descendentibus aut ascendentibus *Iehovae* Exercitibus, ita repletum est; ut nullus inanis locus relinquatur."

51. Lansbergen 1663, 29: "Lux enim unius *Solis,* si tam eximia sit, ut oculi sine damno eam non ferant; quanto fortior censenda est, tot ac tantorum lucidorum Corporum; quae in *Secundo Coelo* colligitur et coit? Recte ergo *Apostolus,* 1 Tim. 6. 16. *Deum* ait *habitare in luce inaccessibili.* Nam si tam illustris sit splenor *Octavae* Sphaerae, quae Atrium est *Divini Palatii;* quam fortis et inaccessus erit Fulgor ipsius *Habitaculi Divinae* Majestatis?"; Graney 2013a, 169.

52. Lansbergen 1663, 30: "*Fixas* porro, *Deus* in *Firmamento* collocavit, ut *Exercitus.* . . . Sunt enim vere, *Exercitus Dei* visibiles; quibus ut *Deus Zebaoth,* hoc est, *Exercituum,* fortissime indies praeliatur"; Graney 2013a, 169. The idea that stars are living beings has a long history. It dates back to at least Origen (see Scott 1991), and was addressed and rejected by Thomas Aquinas.

53. Lansbergen 1663, 30: "Nam sicut *Deus,* est *Deus Exercituum,* qui imperat; ita *Stellae* sunt praeliantes *Dei Exercitus,* qui parent, et Mandata ejus perficiunt."

54. Lansbergen 1663, 30: "Nam unaquaeque non *Terram* tantum, sed et *Solem*, et totum fere *Orbem Terrae* excedit. *Deus* autem sine dubio Vires iis addidit, Magnitudini analogas; ut instar Gigantum, ac fortium Bellatorum, Ordine in Coelo consistant, sine ullo motu. Nam uti nemo *Terram* movere potest, nisi solus *Deus* qui eam condidit, Hagg. 2. 22. sic nemo immensa ac praevalida haec Corpora, commovere novit; nisi *Deus* eorum Creator. Sunt ergo, vere magni et potentes *Dei Exercitus*, per quos, *efficit quicquid vult* Psal. 115. 3"; Graney 2013a, 169.

55. Lansbergen 1663, 31: "Patet ergo denuo, *Exercitum Fixarum*, non magnum solummodo; sed et omnibus, qui unquam fuere in Terris, majorem esse: ut jure merito vocentur *Exercitus Domini*."

56. Lansbergen 1663, 31.

57. Lansbergen 1663, 31.

Chapter 6. Jesuits on the Tower

1. Huygens 1722, 145.

2. Koyré 1953, 231–32; Koyré 1955, 349; Lindberg and Numbers 1986, 155; Heilbron 1999, 180.

3. Grant 1984, 12.

4. Koyré 1953, 230.

5. Koyré 1953, 230.

6. In his *The Sun in the Church: Cathedrals as Solar Observatories*, J. L. Heilbron also describes Riccioli as seeking a better pendulum, much as Koyré does here. Heilbron speaks of Riccioli settling on a final length "which—such is the confidence of faith—he accepted without trying" (Heilbron 1999, 180). My reading of Riccioli's *New Almagest* discussion of these pendulums is of a dry discussion of results obtained via different methods, with little sense of disaster, disappointment, or confidence of faith.

7. Riccioli 1651, 1:86: "quia vmbrae ad lineas horarias appulsus non potest discerni adeo exacte, vt non formidemus de aliquorum Secundorum errore."

8. See Graney 2012d.

9. Riccioli 1651, 2:384. See appendix B, part 2, for the Latin for this quote and the other quotes from Riccioli's writings on falling bodies.

10. Riccioli 1651, 2:384. Recall that the concept of impetus was discussed in chapter 2 of this volume.

11. Aeschylus was an ancient Greek tragedian who, according to legend, was killed when an eagle mistook his bald head for a stone and dropped a tortoise on it in an attempt to break the tortoise's shell.

12. One of Odysseus's crew.

13. The Ovid translation is from Wheeler 1939, 117.
14. Riccioli 1651, 2:384.
15. Riccioli 1651, 2:384–85.
16. Riccioli 1651, 2:385.
17. Riccioli 1651, 2:385.
18. Riccioli 1651, 2:385.
19. Riccioli 1651, 2:385.
20. Riccioli 1651, 2:385.
21. Riccioli 1651, 2:385–86.
22. See Galilei 2001, 257.
23. See Galilei 2001, 259.
24. Riccioli 1651, 2:386.
25. Riccioli 1651, 2:386.
26. Riccioli 1651, 2:386.
27. Riccioli 1651, 2:386.
28. Riccioli 1651, 2:386.
29. Riccioli 1651, 2:387.
30. Riccioli 1651, 2:388.
31. Riccioli 1651, 2:388.
32. Riccioli 1651, 2:389.
33. Riccioli 1651, 2:389.
34. Meli 2006, 131–34.

Chapter 7. 126 Arguments

1. Grant 1996, 652.
2. In fact Riccioli advocated a variation on Tycho's hypothesis.
3. Playfair 1824, 97.
4. Figuier 1870, 89.
5. Van Helden 1989, 103.
6. Grant 1996, 63.
7. Dinis 2003, 209.
8. Linton 2004, 226–27.
9. Another inaccurate assessment of Riccioli is worth mention. I have consulted Maurice Finocchiaro's work extensively, but even he has mischaracterized Riccioli. Finocchiaro writes: "According to Riccioli, the literal meaning of biblical statements is physically true and scientifically (philosophically) correct, and it takes precedence in cases of conflict with doctrines in natural philosophy. One interesting consideration he advanced is the following holistic argument: 'If the liberty taken by the Copernicans to interpret scriptural texts and to elude ecclesiastic

decrees is tolerated, then one would have to fear that it would not be limited to astronomy and natural philosophy and that it could extend to the most holy dogmas; thus it is important to maintain the rule of interpreting all sacred texts in their literal sense'" (Finocchiaro 2005, 83–84). Unfortunately, he cites this as "Riccioli 1651, 2:290; here quoted and translated from the French text in Delambre 1821, 1:672" (Finocchiaro 2005, 383, n. 111; both sources are in the bibliography of this volume). What Riccioli in fact writes is "Enimuero si Copernicanis eam, quam sibi assumpsere licentiam interpretandi diuinas literas, et Ecclesiastica decreta eludendi, concesserimus, ea non intra fines Astronomiae solius, aut Philosophiae naturalis fortasse continebitur, sed ad alia quoque sanctiora dogmata per alios extendi poterit; si nimirum semel *absque manifesta necessitate* literalem sensum diuini codicis abnegare licuerit," which would change Finocchiaro's text to read: "... thus, *except in cases of manifest necessity*, it is important to maintain the rule of interpreting all sacred texts in their literal sense." I have added the italics. The "manifest necessity" is not in Delambre. Riccioli's next sentence is that there is no necessity of retreating from the literal sense on account of physics or astronomy, as he will show—through his analysis of 126 arguments.

10. Riccioli 1651, 2:475 (par. 27 and *Responsum*): "quia aues, naues etc. non solo motu proprio, sed communi etiam pariter cum Terra mouentur in Copernici hypothesi."

11. Riccioli 1651, 2:476 (par. 60).

12. Riccioli 1651, 2:476 (par. 61).

13. Riccioli 1651, 2:468 (par. 15): "Si Fixae mouerentur diurno motu potius quam Tellus; aliae non modo diuerso, sed eodem quoque tempore describerent circulos maximos, nempe illae; quae in Aequatore; aliae minimos, nempe que prope polos, et sic illae velocissimae, hae tardissimae essent. At hoc est absurdum. Ergo Tellus potius motu diurno mouetur, quam Fixae."

14. Riccioli 1651, 2:468 (par. 15, *Responsum*).

15. Galilei 2001, 138–39.

16. Riccioli 1651, 2:468 (par. 15): "Huius quoque argumenti prolati me puderet, nisi totus pudor redundaret in Galilaeum, qui serio illud vrget, vapulaturus vtique a Tyrunculis Astronomiae."

17. Galilei 2001, 138.

18. Riccioli 1651, 2:468 (par. 12, *Responsum*).

19. Riccioli 1651, 2:466 (par. 6): "Diurnus motus tribuendus est illi potius corpori, quod certo constat esse sic mobile, quam ei de cuius mobilitate non sumus certi. Atqui de mobilitate Telluris sumus certi, quia certi sumus de ipsius finitate; de cæli autem supremi mobilitate tam incerti sumus, quam incertum est sitne finitum an infinitum: nam si infinitum esset aut non esset mobile reuolutione diurna, aut saltem controuersum est inter Physicos an esset mobile."

20. Riccioli 1651, 2:466 (par. 6, *Responsum*).

21. Riccioli 1651, 2:475 (par. 39 and *Respondit*).

22. Galilei 2001, 285.

23. Riccioli 1651, 2:475–76 (pars. 39–50).

24. Graney 2012b, 220–22.

25. Galilei 2001, 189–94.

26. Galilei 2001, 193.

27. Galilei 2001, 194.

28. Gingerich 2006, 91–95.

29. Heilbron 1999, 178–81.

30. Riccioli 1651, 2:472 (pars. 49 through 2).

31. Riccioli 1651, 2:471 (par. 43 and *Responsum*).

32. Galilei 2001, 483–539.

33. Riccioli 1651, 2:472 (par. 48): "nullo alio commodiore modo explicantur, et euidentiore, vel saltem probabiliore causa adducta, quam per inæqualem motum Telluris, ortum ex varia permixtione diurni motus cum annuo."

34. Finocchiaro 1989, 35, 273–74.

35. Riccioli 1651, 2:472 (par. 48, *Responsum*): "ad compescendam Galilaei in hoc argumento iactantiam"; "nulla adhuc opinio de causa æstus Maris emicuerit, quæ omnes difficultates tollat, et intellectui omnes æstuum differentias contemplanti satisfaciat; aliquæ tamen sunt, quæ multo probabiliores sunt, quam Galilæi opinio."

36. Riccioli 1651, 2472 (par. 48, *Responsum*).

37. Riccioli 1651, 2:471 (par. 42, *Responsum*): "et tamen nobis nec mane, nec vesperi, nec circa Meridiem nec Medinoctium (quod melius esset, quia tunc differentia, qua motus diurnus annuo adderet, esset euidentior, quam mane aut vespere) differentia vlla certa et sensibilis apparuit inter vnius horæ æqualis ex stellarum transitibus notæ, et alterius vibrationes." As seen in chapter 6, Riccioli precisely calibrated pendulums that would work well for this experiment.

38. Riccioli 1651, 2:470 (par. 32 and *Responsum*): "et quidem absque vlla necessitate aut fundamento a sensibus sumpto, qui nullum motum Terræ nedum inæqualitatem eius perfentiscunt."

39. Riccioli 1651, 2:477 (par. 76 and *Responsum*): "ex vtroque enim æstimanda esset, licet diuerso modo, hæc mensura."

40. Riccioli 1651, 2:477 (par. 77 and *Responsum*).

Chapter 8. An Angel and a Cannon

1. Galilei 2001, 216–18.

2. Galilei 2001, 218.

3. Galilei 2001, 206–12.

4. Galilei 2001, 206–7. Galileo even discusses whether the turning motion of a gun pointed at and following a flying bird might be transmitted to the bullet.

5. Galilei 2001, 211.

6. Riccioli 1651, 2:474 (par. 18 and *Respondit*).

7. Riccioli 1651, 2:425: "si globus exploderetur versus polos per planum eiusdem Meridiani, minor illi diuersitas a motu diurno inferretur, quam si modo versus Ortum, modo versus Occasum; Si vero in parallelis polo propioribus, tardius cum terra, si in propioribus Aequatori celerius cum terra ferretur ille globus, coeteris, vt supponitur, paribus"; Graney 2011b, 390; Graney 2011c. A closer translation might read "if a ball might be ejected toward the poles over level ground of the same Meridian, a difference from the diurnal motion might be brought in (smaller to that, than if [ejected] just toward the East, just toward the West); truly that ball might be carried off, if [over] slower ground with the earth, in the case of parallels nearer to the pole, if [over] faster ground with the earth, in the case of nearer to the Equator, the rest, as is supposed, equal."

8. Riccioli 1651, 2:426–27.

9. Riccioli 1651, 2:474 (pars. 17 and 19).

10. Riccioli 1651, 2:427 (*Si Tellus*); Graney 2011c.

11. Riccioli 1651, 2:473 (par. 10): "Si Angelus sphaeram metallicam magni ponderis catenae appensam dimitteret; catena illa vi sphaerae extenderetur perpendiculariter versus Terram; At secundum Copernicanos deberet oblique curuari versum Orientem"; Graney 2011b, 390–91.

12. Van Zwoll 2011, sec. 4. Van Zwoll states, "Here's a rule of thumb for hitting distant targets with a sporting rifle: Figure a 1 inch Coriolis correction for each second of bullet flight time." Since its cause is the rotation of the Earth, the Coriolis effect is also connected to the vertical deflection of projectiles launched east or west that Galileo described, which is sometimes called the "Eötvös effect," after the Hungarian physicist Loránd Eötvös, who detected it in the early twentieth century. Interestingly, Galileo agrees with Van Zwoll in expecting a deflection of one inch for a shot that travels through the air for one second (Galilei 2001, 211). A commonly cited example of the Coriolis effect manifesting itself more dramatically, in projectiles hurled over longer ranges, is in the Falklands naval battle of the First World War, where a failure to correctly account for it supposedly caused British gunners to initially miss their targets by one hundred yards or so. See Littlewood 1953, 51; Marion 1970, 343–56; Beatty 2006, 194–95. The Coriolis effect on projectiles launched from the German "Paris Gun," which bombarded Paris from a range of seventy-five miles during the First World War, was sufficient to cause a deflection of several hundred yards (Denny 2011, 124).

13. Riccioli 1651, 2:474 (par. 17): "Sed profecto fieri possunt: nec discrimen inter praedicata interualla futurum est insensibile, cum adeo violentus sit vterque motus."

14. If the tower is of height h, and the Earth has radius R and makes one rotation in time T, then the speed of the bottom of the tower is the Earth's circumference divided by time, or $v = 2\pi R/T$, while the top of the tower exceeds this speed by a fraction equal to h/R. Thus, the top of the tower exceeds the bottom by speed $s = (2\pi R/T)(h/R) = 2\pi h/T$. The time required for a heavy body to fall from the top of the tower can be found from kinematic equations: $t = (2h/g)^{1/2}$. As we have seen with Riccioli's falling bodies experiments, this time can also be directly measured. The eastward deflection of the ball then is $d = st = (2\pi/T)(2h^3/g)^{1/2}$ (Graney 2011b, 391). Thus, an object falling at the equator from a tower 100 meters high (for example, from the Asinelli tower in Bologna, which Riccioli used in his experiments) should be deflected eastward by approximately 3.3 centimeters, or a little over an inch; if it were at the poles, it would not be deflected at all. This is a rough calculation—a modern calculation of the Coriolis force gives an eastward deflection of $d \approx (2/3)(2\pi/T)\cos(\lambda)(2h^3/g)^{1/2}$, where λ is the latitude (Marion 1970, 350). This formula yields an eastward deflection of 2.2 cm at the equator, 1.6 cm at 45° N latitude (Bologna), and zero at the poles.

15. Koyré 1955, 374.

16. Hooke 1674, 5.

17. Ball 1893, 142–43.

18. Ball 1893, 145. The reader who might find questionable Riccioli's assertion that skilled gunners could place a ball right into the mouth of an enemy cannon will no doubt be interested to learn that the minutes next show that Sir Christopher Wren proposed a variation of this experiment, in which a gun would be fired upward, after which follows: "Mr. Flamstead hereupon alledged, that it was an observation of the gunners, that to make a ball fall into the mouth of the piece, it must be shot at eighty-seven degrees; and that he knew the reason thereof; and that it agreed with his theory: and that a ball shot perpendicularly would never fall perpendicularly." Extreme precision in gunnery, to the extent that gunners could drop cannon balls, not merely into the mouths of opponents' cannons, but right back into the mouths of their own cannons, seems to have been taken for granted in the mid-seventeenth century.

19. Ball 1893, 148.

20. Ball 1893, 149.

21. Ball 1893, 150.

22. Hall 1903, 182.

23. A fair question is whether Riccioli ever performed an experiment to detect for the existence or nonexistence of eastward deflection. He does not make mention of such an experiment in connection with his anti-Copernican argu-

ment number 10. It is possible I have overlooked such a discussion in the vastness of the *New Almagest*. However, Riccioli's falling body experiments all involved the basics of what Newton proposed and Hooke carried out—dropping bodies to a point indicated by a plumb line (Riccioli describes the use of plumb lines in his falling bodies experiment), and he was familiar with Tycho's arguments for a westward deflection in falling bodies. I speculate that he would have watched for anything unusual in the impact points of his falling bodies, and that he would have been of the opinion that, if anything unexpected was occurring in that regard, he would have noticed it.

24. Hall 1903, 182–83; Meli 1992, 427–30.

25. Rigge 1913, 209–10.

26. Hall 1903, 189; Rigge 1913, 209–10.

27. Rigge 1913, 210; Marion 1970, 350 (footnote); Hall 1903, 186–88; Meli 1992, 446.

28. Hall 1903, 188–90; Meli 1992, 446; Cajori 1901.

29. Akmaev 2012.

30. See, for example, Gregersen 2011, 121.

31. Riccioli 1651, 2:474 (par. 19).

32. Riccioli imagined a way to get around this problem of limited fall heights: perform the experiment on the deflection of a falling body in reverse! Recall from chapter 2 that under Aristotelian physics, objects composed of elemental fire had "levity"—they rose from the center of the universe (and of the Earth) just as heavy bodies fell toward that center. Thus Riccioli imagined that, were it possible to find a body that was not composed of a mixture of elements (as bubbles or earthly fire or other things that rise up in the air are composed), but rather was composed of pure elemental fire, then it would be the reverse of a heavy falling body. Whereas a heavy ball would drop vertically to the ground under the influence of its gravity if Earth did not rotate, or would fall with a deflection to the east if Earth did rotate, a body of pure fire would ascend vertically upward under the influence of its levity if Earth did not rotate, or would ascend with a deflection to the *west* if Earth did rotate. In such an experiment, there would be no limit imposed by the height of available towers—only by the limit of what the eye or telescope could see. See Riccioli 1651, 2:417.

Chapter 9. The Telescope against Copernicus

1. Riccioli 1651, 1:715; Graney 2010b, 457.

2. Lansbergen 1631, 130–34.

3. Riccioli 1651, 1:715–16: "denudanti discos stellarum, et abradenti cincinnos radiorum aduentitios"; Graney 2010b, 457–58.

4. Riccioli 1651, 1:715–16; Graney 2010b, 457–58. Elsewhere Riccioli also discusses Hortensius and Galileo, arguing that just as Tycho and others overestimated the diameters of the stars, Hortensius and Galileo underestimate them, while noting that he and Grimaldi offer diameters measured with certainty by the method described here (Riccioli 1651, 2:460–61).

5. Riccioli 1651, 1:716; Graney 2010b, 459.

6. Riccioli 1651, 1:716; Graney 2010b, 459–61.

7. Graney 2010b, 459–61. Riccioli's book is plagued with typographical errors, so that often semidiameters are stated as diameters, and so forth (Graney 2010b, 461, 465). The values given here are corrected, not necessarily the values Riccioli presents. However, usually the numbers involved are so large than a factor of two error does not change the overall picture very much.

8. Hortensius 1633, 60–64; Graney 2010b, 462.

9. Hortensius makes an effort to solve the problem by suggesting that a parallax of 60 seconds is still plausible, in which case the stars can be correspondingly closer to Earth and thus correspondingly smaller (Graney 2010b, 462).

10. For example, Riccioli 1651, 2:475 (par. 67, *Responsum*; par. 70, *Respondent*).

11. Riccioli 1651, 2:475 (par. 70, *Respondent*): "etsi falsitatis redargui non possit; prudentioribus tamen viris non posse satisfacere."

12. Much of the following discussion is adapted from Graney 2012a, 113–15.

13. Riccioli 1651, 2:462.

14. Riccioli 1651, 2:462: "vel certe aliquo alio indicio sensibili nobis permisisset venire in certam notitiam huius distantiae ac magnitudinis; id vero nequaquam fecisse, patet ex hactenus dictis, cum omnia phaenomena Astronomica salua sint sine Copernicaea hypothesi; physica vero experimenta grauium, et percussiones proiectorum euidenter illam hypothesim falsitatis redarguant."

15. Riccioli only gives a short phrase, "exultasse ut gigantem ad curredam viam" (which can be compared to "exultavit ut gigans ad currendam viam suam" from the Vulgate Psalm 18:6). I have included a longer quote from the Douay-Rheims version to provide context. In modern versions of the Bible this is Psalm 19. See Riccioli 1651, 2:462 (bottom of second column).

16. Riccioli gives the short phrase, "vas admirabile opus excelsi" (matching the Vulgate Ecclesiasticus 43:2). The quote is from the Douay-Rheims version of Ecclesiasticus 43:1–3a. See Riccioli 1651, 2:462–63.

17. Riccioli 1651, 2:462–63.

18. Riccioli 1651, 2:462, 467 (par. 11 and *Responsum*).

19. This is a paraphrasing of sentiments expressed in various places in the *New Almagest*, such as Riccioli 1651, 2:462, 467 (par. 9, *Responsum*), 477 (par. 70, *Respondent*).

Chapter 10. It Can No Longer Be Called "False and Absurd"

1. Rubenstein 2003, x.
2. Galilei 2001, back cover.
3. Bronowski 2011, 156.
4. Dinis 2003, 209.
5. Magruder 2009, 208–9.
6. Graney and Sipes 2009.
7. In 1662 Johannes Hevelius published a table of star diameters measured with such accuracy that they could serve as useful data for modern astronomers. His results were that the diameters of first magnitude stars measured a little more than six seconds (Graney 2009).
8. The reader may at this point wonder whether Kepler's work would not have been on the side of the Copernicans in this imaginary scenario. The answer is "no." Riccioli discussed Kepler's work in the *New Almagest*. The accuracy of Kepler's elliptical orbits supported neither side in the debate, for, as Riccioli noted, the Sun could orbit the Earth with an elliptical orbit just as well as the Earth could orbit the Sun—see Riccioli 1651, 2:468 (par. 17). Indeed, Riccioli would go on to fully adopt elliptical orbits into the Tychonic hypothesis—see Wilson 1989, 185. Another supporter of the Tychonic system who incorporated elliptical orbits into that system was Jean Morin (1591–1659); see Schofield 1989, 42.
9. Delambre 1821, 678: "Diamètres, mouvemens et distances des fixes. Rien de certain de part ni d'autre."
10. Walsh 1969, 200. This same assessment also appears in older versions of the *Catholic Encyclopedia*, 1913; see 13:40. Thanks to the online version of that encyclopedia, http://www.newadvent.org/cathen/13040a.htm, it is present in many places on the Internet.
11. *New Catholic Encyclopedia* 2003.
12. Wootton (2010, 260–63) argues that Galileo sabotaged that opportunity by betraying Urban, his ally and friend.
13. I owe this turn of phrase to a blogger: M. Francis, "The Book Galileo Was Supposed to Write," 21 April 2011, http://m-francis.livejournal.com/196433.html.
14. Graney 2012a, 118–19.
15. See, for example, "Paul Offit and 'Deadly Choices,'" *Science Friday*, 7 January 2011, available online at www.sciencefriday.com/program/archives/201101075.
16. Graney 2012c, 19–20.
17. Applebaum 2012, 63.
18. Applebaum 2012, 64.
19. Applebaum 2012, 58–72; Graney 2013b, 119.

20. Huygens 1722, 145. I thank Dennis Danielson for bringing this passage to my attention.

21. Graney 2010b, 466, note 39.

22. Bailey 1835, 205. Note Flamsteed's reference to "parts," just as Riccioli does in his tables of star diameters.

23. Halley 1720, 1.

24. Halley 1720, 3.

25. For a detailed, technical discussion of diffraction, Airy's theory, and the appearance of stars as seen through early telescopes, see Grayson and Graney 2011.

26. In fact, modern telescope users can try Riccioli's technique for themselves. If a modern user aims his or her telescope at a bright star, and places over the telescope a mask (which can be made of light cardboard or dark construction paper, with nothing directly touching the lens or mirror of the telescope) that contains a hole with aperture of, for example, ¾ of an inch, matching the aperture Flamsteed used, he or she will see stars more or less as Flamsteed saw them. Trying different size holes until no more diffraction rings are seen on the brightest stars—with only a clean disk apparent—will produce for the modern user a telescope that, according to Riccioli, is optimized for viewing stars. Observing the stars with such a telescope is an excellent education in seeing stars the way seventeenth-century observers saw them.

27. As is discussed in the paragraph of the main text that follows this note, reducing aperture enlarges the spurious disk while it dims the diffraction rings. Thus this process can produce what appears to be an optimal adjustment of the telescope, such as is obtained when focusing it. Filtering through smoke does not enlarge the spurious disk; the disk dims and shrinks along with the rings. No optimal point is reached. See Grayson and Graney 2011 for a detailed, technical discussion of the results of masking a telescope's aperture vs. filtering the light.

28. Aberrations in the eye also play a role—a person with excellent vision sees stars that are small and round; a person with poor vision sees stars that are large and fuzzy looking.

29. Grayson and Graney 2011.

30. The Earth's motion through light will affect the direction from which the light appears to come. A similar phenomenon is experienced by a woman running in an easterly direction through rain, on a day with no wind. The rain falls straight down, and were the woman standing still, it would hit her on the top of the head. But because the woman is moving, she moves into the rain. The result is that the rain strikes her face. Thus it seems to her to be coming toward her—coming from the east. If she turns around and runs to the west, the rain still strikes her face—and thus now seems to be coming from the west. Stellar aberration is this basic

phenomenon, occurring with light. It causes the direction from which a star's light appears to come, and therefore the position of the star in the sky, to be slightly affected over the course of a year by the changing direction of Earth's motion around the Sun.

31. Finocchiaro 2005, 139.

32. Finocchiaro 2005, 142–43.

33. Finocchiaro 2005, 143.

34. Finocchiaro 2005, 145.

35. Finocchiaro 2005, 145.

36. Finocchiaro 2005, 147.

37. Finocchiaro 2005, 147–48.

38. "Calandrelli, Giuseppe" 2007, 191–192; Wallace 1995, 12–13. Calandrelli was the preeminent astronomer in Rome in the early nineteenth century; his measurements were eventually shown to be erroneous, but during the debate in which Olivieri took part, this work was cited as a demonstration of Earth's motion.

39. Finocchiaro 2005, 211.

40. Finocchiaro 2005, 207.

41. Finocchiaro 2005, 127, 129, 198.

42. "And the angel of the Lord called to Abraham a second time from heaven, saying: By my own self have I sworn, saith the Lord: because thou hast done this thing, and hast not spared thy only begotten son for my sake: I will bless thee, and I will multiply thy seed as the stars of heaven, and as the sand that is by the seashore," Genesis 22:15–17. The stars in the sky and the sand by the seashore are not numerically comparable. If the number of stars in the sky refers to the number of stars visible to the eye of Abraham, of which there are merely some thousands (despite songs about millions of stars in the sky to the contrary), then they are vastly outnumbered by the grains in a cup of sand. If the number of stars in the sky refers to the number of stars in the universe, then they vastly outnumber all the sands on any seashore.

43. Riccioli 1651, 2:478. See note 5 in chapter 1 of this volume for the Latin.

Appendix A. Francesco Ingoli's 1616 Essay to Galileo

1. Giovanni Antonio Magini (1555–1617). See Magini 1589, 104. Ingoli's values contain errors and agree with neither Magini nor the 1:22 ratio he cites in this paragraph.

2. Erasmus Reinhold (1511–1553).

3. "Parallax limits" discussed by Copernicus in *On the Revolutions*, book 4, chapter 22—I thank Eduardo Vila Echagüe for pointing this out to me.

4. Ingoli here is envisioning the heliocentric Sun opposite the Moon in the sky, so that the Sun, Earth, and Moon are aligned in that order. The distance between Sun and Earth is 1179 terrestrial semidiameters. The distance between Earth and Moon is 65. Thus in a heliocentric system, measuring from the Sun through the Earth, and then through the Moon to the orb of the stars, we see that the Moon is $1179 + 65 = 1244$ semidiameters closer to the stars than is the Sun.

5. John of Holywood (ca. 1195–ca. 1256).

6. Ingoli is noting various measurements on the celestial sphere that can be used to show that half of the sky is indeed visible.

7. Favaro has 20 minutes. However, in his reply to Ingoli, Galileo speaks of Ingoli wanting "to make Copernicus's annual orbit imperceptible" (Finocchiaro 1989, 170), and 20 minutes, being two-thirds the size of the full Moon, is not imperceptible. Galileo understood Ingoli to be referencing a small value, so it seems likely 2 is the proper value.

8. Again, Favaro has 20 minutes.

9. Genesis 1:14.

10. Douay-Rheims version. This is Psalm 104:2 in more recent versions, such as the New Revised Standard Version, in which the translation reads, "You stretch out the heavens like a tent."

11. This is Psalm 139:8 in the New Revised Standard Version, in which the translation reads, "If I ascend to heaven, you are there; if I make my bed in Sheol, you are there."

12. Isaiah 14:13, 15.

13. This last sentence is very much a paraphrase. A closer translation would be, "Indeed just as it befalls to the man who makes a journey from south to north or north to south, the altitude of the pole to be changed, thus [it befalls] to the place the altitude of the pole to be changed, if it itself may be moved instead of the man."

14. See Galileo's comments in Finocchiaro 1989, 191.

15. Joshua 10:13–14.

16. Ingoli is speaking of two groups of "Fathers"—the Church Fathers (from ancient times) and the Synod Fathers of the Council of Trent.

17. Taken from Martis 2007, 632. A closer translation reads,

Mighty Builder of the Earth Who pulling up the ground of the World,
The troubles of the waters banished, You have made the Earth immobile.

18. Finocchiaro 1989, 157.

19. Finocchiaro 1989, 162–65.

20. Finocchiaro 1989, 166.

21. Finocchiaro 1989, 174.

22. Finocchiaro 1989, 175–81.

23. Finocchiaro 1989, 180. Galileo may have acquired this idea from Giordano Bruno, who wrote, "there must be a primary body which must be of itself both bright and hot and consequently also unchanging, solid and dense; for a rare and tenuous body cannot hold either light or heat, as we shew elsewhere . . . and the sun by virtue of those parts which are bright and hot must be like a stone, or a most solid incandescent metal; not a fusible metal as lead, bronze, gold or silver, but an infusible; not indeed a glowing iron but that iron which is itself a flame" (Bruno [1584] 1986, 308).

24. Finocchiaro 1989, 187.

25. Finocchiaro 1989, 190.

26. Finocchiaro 1989, 190.

27. Finocchiaro 1989, 191.

28. Finocchiaro 1989, 191.

29. Finocchiaro 1989, 192.

30. Finocchiaro 1989, 193.

31. Finocchiaro 1989, 195.

32. Finocchiaro 1989, 196.

Appendix B. Riccioli's Reports Regarding His Experiments with Falling Bodies

1. See chap. 2.

2. Aeschylus: An ancient Greek tragedian who, according to legend, was killed when an eagle mistook his bald head for a stone and dropped a tortoise on it in an attempt to break the tortoise's shell.

3. One of Odysseus's crew.

4. For the Ovid translation we have used Wheeler 1939, 117.

5. See Galilei 2001, 257.

6. See Galilei 2001, 259.

7. See Galilei 2001, 257.

8. The Jesuits, or the Society of Jesus.

9. There is a typographical or transcription error here. One second is six strokes of Riccioli's pendulum. Thus 23 strokes would be 3 and 5/6 seconds, rather than the 3 and 1/2 seconds stated here. Also, the 15-foot distance gap between the heavier and lighter balls agrees reasonably well with modern calculations, but the 3-stroke time gap does not (simulations using reasonable estimates for the sizes and densities of the balls Riccioli dropped—"palm-wide" being roughly 9 cm, for example—produce a gap of very roughly 15 feet relatively easily, but a 3 stroke time gap cannot be reasonably simulated). Finally, the 3-stroke time gap does not agree with Riccioli's writings a few pages earlier in the *New Almagest*, where he

concludes that a falling clay ball of 8 ounces takes 5 strokes to fall 10 Roman feet from rest (and 26 strokes to fall 280 Roman feet), and that velocity increases linearly with time. Thus he says, "during the first measure it traverses OC, 10 feet; during the second it traverses CQ, 30 feet; third, QR, 50 feet; fourth, RS, 70 feet; and fifth, ST, 90 feet." For the 10-ounce ball to travel just 15 feet in 3 strokes at around the fourth second of fall is inconsistent with these numbers.

10. Note that here the translation deviates significantly from Riccioli's original Latin. Riccioli wrote the five combinations out in paragraph form, and not in a numbered list.

11. See chap. 2 regarding levity vs. gravity.

12. Capra et al. 2010.

13. For a further analysis of Riccioli's data, see Warren 2013.

14. The words he uses, "gravius in specie"—"heavy in species"—are really more like "specific gravity," but we did not translate them as such because Riccioli's conceptions regarding gravity are not modern. To Riccioli, gravity is not a force of attraction between Earth and a ball, but a ball's natural tendency to move toward the center of the universe (see chap. 2). Thus "heaviness" seemed a more apt translation.

Works Cited

Abrams, M. H. ed. 1962. *The Norton Anthology of English Literature: Major Authors Edition.* New York: Norton.

Akmaev, Rashid A. 2012. "Correcting the Coriolis Correlation." *Physics Today* 65 (January): 8.

Apostle, H. G., and L. P. Peterson. 1986. *Aristotle: Selected Works, Translated from Greek into English.* 2nd ed. Grinnell, IA: The Peripatetic Press.

Applebaum, Wilbur, trans. 2012. *Venus Seen on the Sun: The First Observation of a Transit of Venus by Jeremiah Horrocks.* Leiden: Koninklijke Brill.

Bailey, Francis. 1835. *An Account of the Rev'd John Flamsteed, the First Astronomer-Royal: Compiled from His Own Manuscripts, and Other Authentic Documents.* London.

Ball, W. W. R. 1893. *An Essay on Newton's "Principia."* New York: Macmillan.

Barker, Peter. 2004. "How Rothmann Changed His Mind." *Centaurus* 46:41–57.

Beatty, M. F. 2006. *Dynamics: The Analysis of Motion.* Vol. 2. New York: Springer.

Blair, Ann. 1990. "Tycho Brahe's Critique of Copernicus and the Copernican System." *Journal for the History of Ideas* 51:355–77.

Bond, George P. 1848. "An Account of the Nebula in Andromeda." *Memoirs of the American Academy of Arts and Sciences* n.s. 3:75–86.

Brahe, Tycho. 1601. *Epistolarum Astronomicarum Libri.* Nuremberg.

———. (1602) 1915. *Astronomiae Instauratae Progymnasmata, Pars Secunda.* In J. L. E. Dreyer, *Tychonis Brahe Dani Opera Omnia,* vol. 2. Copenhagen.

"Brahé, Tycho." 1836. In *The Penny Cyclopædia of the Society for the Diffusion of Useful Knowledge,* vol. 5. London.

Bronowski, Jacob. 2011. *The Ascent of Man.* New York: Random House.

Bruno, Giordano. (1584) 1986. *On the Infinite Universe and Worlds*. In Dorothea Waley Singer, *Giordano Bruno: His Life and Thought, with Annotated Translation of His Work "On the Infinite Universe and Worlds."* New York: Greenwood Press.

Buridan, John. 1974. "The Impetus Theory of Projectile Motion." Translated by Marshall Clagett in *A Source Book in Medieval Science*, edited by Edward Grant, 275–80. Cambridge, MA: Harvard University Press.

Cajori, Florian. 1901. "The Unexplained Southerly Deviation of Falling Bodies." *Science* 14 (29 November): 853–55.

"Calandrelli, Giuseppe." 2007. In *Biographical Encyclopedia of Astronomers*. New York: Springer.

Capra, A., E. Bertacchini, E. Boni, C. Castagnetti, and M. Dubbini. 2010. "Terrestrial Laser Scanner for Surveying and Monitoring Middle Age Towers." Presentation at XXIV FIG International Congress (4445). Available at http://www .fig.net/pub/fig2010/ppt/ts04d/ts04d_capra_bertacchini_et_al_ppt_4445.pdf.

The Catholic Encyclopedia. 1913. 15 vols. New York: Encyclopedia Press.

Christianson, John Robert. 2000. *On Tycho's Island: Tycho Brahe and His Assistants, 1570–1601*. Cambridge: Cambridge University Press.

Copernicus, Nicolaus. (1543) 2001. *De Revolutionibus Orbium Coelestium*. Translated by Edward Rosen, in Michael J. Crowe, *Theories of the World from Antiquity to the Copernican Revolution*, 100–133. Mineola, NY: Dover.

Costanzi, Enrico. 1897. *La Chiesa e le Dottrine Copernicane: Note e Considerazioni Storiche*. Sienna: Biblioceta del Clero.

Couper, H., N. Henbest, and A. C. Clarke. 2007. *The History of Astronomy*. Richmond Hill, ON: Firefly Books.

Crüger, Peter. 1631. *Cupediae Astrosophicae*. Breslau.

Danielson, Dennis Richard. 2000. *The Book of the Cosmos: Imagining the Universe from Heraclitus to Hawking*. Cambridge, MA: Perseus Books.

Delambre, J. B. J. 1821. *Historie de L'Astronomie Moderne*. Paris.

Denny, Mark. 2011. *Their Arrows Will Darken the Sun: The Evolution and Science of Ballistics*. Baltimore: Johns Hopkins University Press.

Digges, Thomas. 1573. *Alæ seu Scalæ Mathematicæ*. London.

Dinis, Alfredo. 2003. "Giovanni Battista Riccioli and the Science of His Time." In *Jesuit Science and the Republic of Letters*, edited by Mordechai Feingold, 195–224. Cambridge, MA: MIT Press.

Donahue, William. 1981. *The Dissolution of the Celestial Spheres, 1595–1650*. New York: Arno Press.

Drake, Stillman. 1957. *Discoveries and Opinions of Galileo*. Garden City, NY: Doubleday Anchor Books.

Dreyer, J. L. E. 1890. *Tycho Brahe: A Picture of Scientific Life and Work in the Sixteenth Century*. Edinburgh.

———. 1909. "The Tercentenary of the Telescope." *Nature* 82 (16 December): 190–91.

Fantoli, Annibale. 1994. *Galileo: For Copernicanism and for the Church.* Rome: Vatican Observatory Publications.

Favaro, Antonio, ed. 1890–1909. *Le Opere di Galileo Galilei: Edizione Nazionale Sotto gli Auspicii di Sua Maestà il re d'Italia.* 20 vols. Firenze.

Figuier, L. 1870. *Earth and Sea.* Translated by W. H. D. Adams. London.

Finocchiaro, Maurice A. 1989. *The Galileo Affair: A Documentary History.* Berkeley: University of California Press.

———. 2005. *Retrying Galileo, 1633–1992.* Berkeley: University of California Press.

———. 2010. *Defending Copernicus and Galileo: Critical Reasoning in the Two Affairs.* Dordrecht: Springer.

———. 2014. *The Routledge Guidebook to Galileo's Dialogue.* London: Routledge.

Freedberg, David. 2002. *The Eye of the Lynx: Galileo, His Friends, and the Beginnings of Modern Natural History.* Chicago: University of Chicago Press.

Galilei, Galileo. 1623. *Il Saggiatore.* Rome.

———. 1989. "Galileo's Reply to Ingoli [1624]." In Finocchiaro 1989, 154–97.

———. 2001. *Dialogue Concerning the Two Chief World Systems, Ptolemaic and Copernican.* Translated by Stillman Drake. New York: The Modern Library.

Gingerich, Owen. 1973. "Copernicus and Tycho." *Scientific American* 229 (6): 86–101.

———. 1993. *The Eye of Heaven: Ptolemy, Copernicus, Kepler.* New York: American Institute of Physics.

———. 2006. *God's Universe.* Cambridge, MA: Harvard University Press.

———. 2009. "Galileo Opens the Door," Harvard-Smithsonian Center for Astrophysics 2009 Observatory Night Video Archive, 19 February 2009. Available at http://www.cfa.harvard.edu/events/mon_video_archive09.html.

Gingerich, Owen, and James R. Voelkel. 1998. "Tycho Brahe's Copernican Campaign." *Journal for the History of Astronomy* 29:1–34.

Goulding, Robert. 2006. "Wings (or Stairs) to the Heavens: The Parallactic Treatises of John Dee and Thomas Digges." In *John Dee: Interdisciplinary Studies in English Renaissance Thought,* edited by Stephen Clucas, 41–63. Dordrect: Springer.

Graney, Christopher M. 2007. "On the Accuracy of Galileo's Observations." *Baltic Astronomy* 15:443–49.

———. 2008. "But Still, It Moves: Tides, Stellar Parallax, and Galileo's Commitment to the Copernican Theory." *Physics in Perspective* 10:258–68.

———. 2009. "Seventeenth-Century Photometric Data in the Form of Telescopic Measurements of the Apparent Diameters of Stars by Johannes Hevelius." *Baltic Astronomy* 17:253–63.

———. 2010a. "Seeds of a Tychonic Revolution: Telescopic Observations of the Stars by Galileo Galilei and Simon Marius." *Physics in Perspective* 12:4–24.

———. 2010b. "The Telescope Against Copernicus: Star Observations by Riccioli Supporting a Geocentric Universe." *Journal for the History of Astronomy* 41:453–67.

———. 2010c. "Changes in the Cloud Belts of Jupiter, 1630–1664, as Reported in the 1665 *Astronomia Reformata* of Giovanni Battista Riccioli." *Baltic Astronomy* 19:265–71.

———. 2011a. "A True Demonstration: Bellarmine and the Stars as Evidence against Earth's Motion in the Early Seventeenth Century." *Logos: A Journal of Catholic Thought and Culture* 14:69–85.

———. 2011b. "Contra Galileo: Riccioli's 'Coriolis-Force' Argument on the Earth's Diurnal Rotation." *Physics in Perspective* 13:387–400.

———. 2011c. "Coriolis Effect, Two Centuries before Coriolis." *Physics Today* 64:8.

———. 2011d. "126 Arguments Concerning the Motion of the Earth, as Presented by Giovanni Battista Riccioli in His 1651 *Almagestum Novum*." arXiv:1103.2057. Available at http://arxiv.org/abs/1103.2057.

———. 2012a. "The Work of the Best and Greatest Artist: A Forgotten Story of Religion, Science, and Stars in the Copernican Revolution." *Logos: A Journal of Catholic Thought and Culture* 15:97–124.

———. 2012b. "Science Rather Than God: Riccioli's Review of the Case for and against the Copernican Hypothesis." *Journal for the History of Astronomy* 43:215–25.

———. 2012c. "Teaching Galileo? Get to Know Riccioli! What a Forgotten Italian Astronomer Can Teach Students about How Science Works." *The Physics Teacher* 50:18–21.

———. 2012d. "Anatomy of a Fall: Giovanni Battista Riccioli and the Story of g." *Physics Today* 65 (September): 36–40.

———. 2013a. "Stars as the Armies of God: Lansbergen's Incorporation of Tycho Brahe's Star-Size Argument into the Copernican Theory." *Journal for the History of Astronomy* 44:165–72.

———. 2013b. "Horrocks on the Transit of Venus." *Journal for the History of Astronomy* 44:119–21.

———. 2013c. "Mass, Speed, Direction: John Buridan's Fourteenth-Century Concept of Momentum." *The Physics Teacher* 51:411–14.

———. 2014. "The Inquisition's Semicolon: Punctuation, Translation, and Science in the 1616 Condemnation of the Copernican System." arXiv:1402.6168. Available at http://arxiv.org/abs/1402.6168.

Graney, Christopher M., and H. Sipes. 2009. "Regarding the Potential Impact of Double Star Observations on Conceptions of the Universe of Stars in the Early Seventeenth Century." *Baltic Astronomy* 18:93–108.

Grant, Edward. 1984. "In Defense of the Earth's Centrality and Immobility: Scholastic Reaction to Copernicanism in the Seventeenth Century." *Transactions of the American Philosophical Society* n.s. 74:1–69.

———. 1996. *Planets, Stars, and Orbs: The Medieval Cosmos, 1200–1687.* Cambridge: Cambridge University Press.

Grayson, T. P., and Christopher M. Graney. 2011. "On the Telescopic Disks of Stars: A Review and Analysis of Stellar Observations from the Early Seventeenth through the Middle Nineteenth Centuries." *Annals of Science* 68:351–73.

Gregersen, E., ed. 2011. *The Britannica Guide to Heat, Force, and Motion.* New York: Britannica Educational.

Grisar, Hartmann. 1882. *Galileistudien, Historisch-Theologische Untersuchungen.* Regensburg: Friedrich Pustet.

Hall, Edwin H. 1903. "Do Falling Bodies Move South?" *The Physical Review* 17:179–90.

Halley, Edmund. 1720. "Some Remarks on a Late Essay of Mr. Cassini, Wherein He Proposes to Find, by Observation, the Parallax and Magnitude of Sirius." *Philosophical Transactions* 31:1–4.

Heilbron, J. L. 1999. *The Sun in the Church: Cathedrals as Solar Observatories.* Cambridge, MA: Harvard University Press.

Herschel, J. F. W. 1828. *Treatises on Physical Astronomy, Light and Sound Contributed to the Encyclopædia Metropolitana.* London and Glasgow: Griffin. Available at http://www.archive.org/details/treatisesonphysi00hersrich.

Hevelius, J. 1673. *Machinae Coelestis pars prior.* Gedani.

Hill, John. 1754. *Urania: or, a Compleat View of the Heavens; Containing the Antient and Modern Astronomy in Form of a Dictionary.* London.

Hooke, Robert. 1674. *An Attempt to Prove the Motion of the Earth by Observations.* London.

Hortensius, M. 1633. *Dissertatio de Mercurio in Sole viso.* Leiden.

Howell, Kenneth J. 2002. *God's Two Books: Copernican Cosmology and Biblical Interpretation in Early Modern Science.* Notre Dame, IN: University of Notre Dame Press.

Huygens, Christiaan. 1722. *The Celestial Worlds Discover'd: or, Conjectures Concerning the Inhabitants, Plants and Productions of the Worlds in the Planets.* 2nd ed. London.

Jarrell, Richard A. 1989. "Contemporaries of Tycho Brahe." In Taton and Wilson 1989, 22–32.

Johnson, Francis R., and Sanford V. Larkey. 1934. "Thomas Digges, the Copernican System, and the Idea of the Infinity of the Universe in 1576." *The Huntington Library Bulletin* 5:69–117.

Jorink, Eric. 2010. *Reading the Book of Nature in the Dutch Golden Age, 1575–1715.* Leiden: Brill.

Keill, John. 1739. *An Introduction to the True Astronomy, or, Astronomical Lectures, Read in the Astronomical School of the University of Oxford.* 3rd ed. London.

Kepler, Johannes. 1995. *Epitome of Copernican Astronomy and Harmonies of the World.* Translated by C. G. Wallis. Amherst, NY: Prometheus Books.

Koyré, A. 1953. "An Experiment in Measurement." *Proceedings of the American Philosophical Society* 97:222–37.

———. 1955. "A Documentary History of the Problem of Fall from Kepler to Newton: De Motu Gravium Naturaliter Cadentium in Hypothesi Terrae Motae." *Transactions of the American Philosophical Society* n.s. 45:329–95.

Lansbergen, Philips. 1631. *Uranometriae libri tres.* Middleburg.

———. 1663. *Opera Omnia.* Middleburg.

Lindberg, D. C., and R. L. Numbers, eds. 1986. *God and Nature: Historical Essays on the Encounter between Christianity and Science.* Berkeley: University of California Press.

Linton, Christopher M. 2004. *From Eudoxus to Einstein: A History of Mathematical Astronomy.* Cambridge: Cambridge University Press.

Littlewood, J. E. 1953. *A Mathematician's Miscellany.* London: Methuen.

Locher, Johann Georg. 1614. *Disquisitiones Mathematicae.* Ingolstadt.

Lohne, J. A.1968. "The Increasing Corruption of Newton's Diagrams." *History of Science* 6:69–89.

Maeyama, Yas. 2002. "Tycho Brahe's Stellar Observations: An Accuracy Test." In *Tycho Brahe and Prague: Crossroads of European Science,* edited by J. R. Christianson, A. Hadravová, P. Hadrava, and M. Šolc, 105–20. Frankfurt: Verlag Harri Deutsch.

Magini, Giovanni Antonio. 1589. *Novae Coelestium Orbium Theoricae.* Venice.

Magruder, Kerry V. 2009. "Jesuit Science After Galileo: The Cosmology of Gabriele Beati." *Centaurus* 51:189–212.

Marion, J. B. 1970. *Classical Dynamics of Particles and Systems.* 2nd ed. New York: Academic Press.

Marius, Simon. 1614. *Mundus Iovialis.* Nuremberg.

———. 1916. "The 'Mundus Jovialis' of Simon Marius, Translated by A. O. Prickard." *The Observatory, A Monthly Review of Astronomy* 39:367–81, 403–12, 443–52, 498–503.

Martis, Douglas. 2007. *The Mundelein Psalter.* Chicago: Liturgy Training Publications.

Meli, D. B. 1992. "St. Peter and the Rotation of the Earth: The Problem of Fall Around 1800." In *The Investigation of Difficult Things,* edited by P. M. Harmon and A. E. Shapiro, 421–47. Cambridge: Cambridge University Press.

———. 2006. *Thinking with Objects: The Transformation of Mechanics in the Seventeenth Century.* Baltimore: Johns Hopkins University Press.

Moesgaard, Kristian Peder. 1972. "Copernican Influence on Tycho Brahe." In *The Reception of Copernicus' Heliocentric Theory*, edited by Jerzy Dobrzycki, 30–55. Boston: D. Reidel.

The New Catholic Encyclopedia. 2003. 2nd ed. Detroit: Thomson Gale/Catholic University of America.

Ondra, Leos. 2004. "A New View of Mizar." *Sky & Telescope* 108 (July): 72–75.

Pagano, Sergio. 2009. *I Documenti Vaticani del Processo di Galileo Galilei (1611–1741)*. New edition. Vatican City: Archivio Segreto Vaticano.

Pannekoek, A. 1961. *A History of Astronomy*. New York: Interscience Publishers.

Playfair, J. 1824. "Progress of Mathematical and Physical Science." In *Encyclopaedia Britannica, supplement to the 4th, 5th, and 6th editions with preliminary dissertations on the history of the sciences*, 1–127. Edinburgh.

Riccioli, Giovanni Battista. 1651. *Almagestum Novum*. 2 vols. Bologna.

Rigge, William F. 1913. "Experimental Proofs of the Earth's Rotation." *Popular Astronomy* 21:208–16.

Rubenstein, Richard E. 2003. *Aristotle's Children*. Orlando: Harcourt.

Schofield, Christine. 1989. "The Tychonic and Semi-Tychonic World Systems." In Taton and Wilson 1989, 33–44.

Scott, Alan. 1991. *Origen and the Life of the Stars: A History of an Idea*. Oxford: Clarendon Press.

Siebert, Harald. 2005. "The Early Search for Stellar Parallax: Galileo, Castelli, and Ramponi." *Journal for the History of Astronomy* 36:251–71.

Tartaglia, Nicolo. 1554. *Quesiti et Inventioni Diverse de Nicolo Tartaglia*. Venice.

Taton, Reni, and Curtis Wilson, eds. 1989. *The General History of Astronomy: Planetary Astronomy from the Renaissance to the Rise of Astrophysics*. Part A. Cambridge: Cambridge University Press.

Thoren, Victor E. 1990. *The Lord of Uraniborg: A Biography of Tycho Brahe*. Cambridge: Cambridge University Press.

"Tycho Brahe's verdensbillede." 2013. Web page of the Tycho Brahe Museum, available at http://www.tychobrahe.com/DK/varldsbild.html.

Van Helden, Albert. 1985. *Measuring the Universe*. Chicago: University of Chicago Press.

———. 1989. "Galileo, Telescopic Astronomy, and the Copernican System." In Taton and Wilson 1989, 81–105.

Van Zwoll, Wayne. 2011. *Shooter's Bible Guide to Rifle Ballistics*. New York: Skyhorse Publishing.

Vermij, Rienk. 2002. *The Calvinist Copernicans: The Reception of the New Astronomy in the Dutch Republic, 1575–1750*. Amsterdam: Koninklijke Nederlandse Akademie van Wetenschappen.

Von Gebler, Karl. 1877. *Galileo Galilei und die Romische Curie.* Stuttgart: J. G. Cotta.

Wallace, William. 1995. "Galileo's Trial and the Proof of the Earth's Motion." *Catholic Dossier* 1:7–13.

Walsh, James J. 1969. *Catholic Churchmen in Science*. Philadelphia: Dolphin Press. First published 1909.

Warren, Patrick. 2013. "Revisiting Riccioli's Free-fall calculations." *Physics Today* 66 (March): 8. Longer version at arXiv:1303.5554; available at http://arxiv .org/abs/1303.5554.

Watson, Fred. 2005. *Stargazer: The Life and Times of the Telescope*. Cambridge, MA: Da Capo Press.

Wertheim, Margaret. 2003. "A Star-Crossed Partnership." A review of Kitty Ferguson, *Tycho and Kepler: The Unlikely Partnership That Forever Changed Our Understanding of the Heavens*. *Los Angeles Times*, May 4. Available at http:// articles.latimes.com/2003/may/04/books/bk-wertheim4.

Wheeler, A. L. 1939. *Ovid with an English Translation: Tristia, Ex Pronto*. Cambridge, MA: Harvard University Press.

Wilson, Curtis. 1989. "Predictive Astronomy in the Century After Kepler." In Taton and Wilson 1989, 161–206.

Wootton, David. 2010. *Galileo: Watcher of the Skies*. New Haven: Yale University Press.

Index

Technical terms in this index appear in bold type. Page numbers in bold indicate where a definition of the term appears in the text or notes.

CHRISTOPHER M. GRANEY

is professor of physics at Jefferson Community & Technical College.

Milton Keynes UK
Ingram Content Group UK Ltd.
UKHW022353261123
433116UK00024B/305